From Passive to Active: How Technology Prevents Accidents

Sharlin

Contents

1 Introduction

Driving a car is a course of action performed by a driver in an environment. During the driving process, the car and the environment, in which the car is conducted, have a reciprocal impact on each other. Driving a car can therefore be regarded as a process with three factors: a driver, a car and the environment. When an accident occurs, one of the factors can be identified as the cause. By analysing the accident protocol, Treat (1977) pointed out that human factors single-handedly or partially account for 93 % of all traffic accidents.

With the ultimate goal of improving traffic safety, we need a systematic understanding of traffic accidents. Kühn and Hannawald (2016) summarize the classification of different phases of traffic accidents which are introduced by European Automobile Manufacturers Association (ACEA). According to that, the development of an accident can be divided into five phases: normal driving, danger, collision unavoidable, during collision and after collision (see Figure 1.1). Passive safety systems are designed to mitigate the impacts of an accident when it happens. Mostly, these systems are activated in the fourth and partly third phases of an accident. Examples of such systems are airbag, seat belt, electronic stability control (ESC) and anti-lock braking system (ABS) (Lie et al., 2004). Moving to the left side of Figure 1.1 is the span before an accident happens (i.e., phase one and two). Assistance systems that are developed for these early phases of an accident are often referred to as active safety systems or advanced driver assistance systems (ADAS). The goal of such systems is to prevent the accident from happening. Examples for ADAS are lane departure warning, blind spot detection, collision avoidance systems, driver drowsiness detection, etc. Ideally, the earlier an assistance system can warn the driver of the potential threat the more time the driver will have to react. However it also poses a challenge since wrong detection of threat will lead to false warnings and drivers will eventually mistrust the systems.

The way how drivers participate in the traffic is regulated by the traffic rules but this by no mean forces everyone behave exactly the same way. The potential occurrence of dangerous situation therefore strongly depends

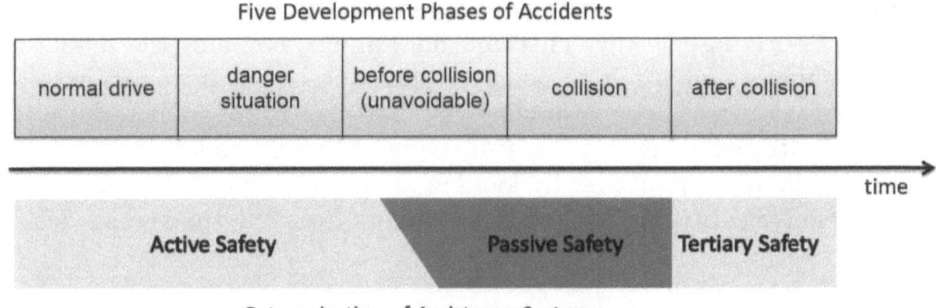

Figure 1.1: Five phases of an accident according to ACEA model and the assignment of safety systems (Kühn and Hannawald, 2016).

on the driver behavior and intention. It leads to the challenge of adapting the driver assistance systems to individual drivers.

Personalization aims at discovering the individual preferences of the driver and use it to improve the prediction of driver's actions or maneuvers. This benefits the safety of advanced driver assistance system since hazardous events can be predicted more accurately. Potential accidents can be better detected and avoided even at the early stages (see Figure 1.1). Furthermore, personalization promises to reduce the number of false warnings and increases the acceptance of drivers.

1.1 Research Questions

As mentioned above, one of the main challenges in developing personalized driver assistance systems is to detect the needs and preferences of individual drivers. In this book, four research questions are addressed in context of personalized advanced driver assistance systems. The answers for these questions are provided and discussed in different aspects throughout the book.

> **Question 1**
>
> How can we learn the driver model from the behavior data to accurately predict the future maneuvers?

Information about the driver is encoded in each decision that is made while she is driving the car. This information also contains the intention for performing a new driving maneuver. The question is if we can extract this information from the behavior data and use it to predict the next maneuver? To what extent is a machine learning model capable of predicting the next maneuver by looking at the past data? Which models are suitable for capturing the driver behaviors from the time series data?

> **Question 2**
>
> How can individual information about the driver be extracted from driving data to further improve the performance of predictive driver assistance systems?

Driving is a complex behavior, containing numerous maneuvers from steering, accelerating, stopping to interacting with other traffic participants. Each maneuver can provide different indicators about the driver e.g. personal traits such as patience or risk-aversion, and also indicators about the driving situations e.g. rush hours. Using various sensors we can record this data in form of a long multivariate time series. The individual information about the driver is hidden in big chunks of data. The challenge is to identify the promising data, extract and transform it into the required format. This will enable us to further improve driver assistance systems.

> **Question 3**
>
> How can we alleviate the problem of intra-personalization to quickly and accurately provide the driver with recommendations?

Intra personalization describes the problem of adapting a system to the temporal changes within the same driver. For example a driver may behave differently when driving to work as when going for vacation. The challenge is therefore to detect the changes in driving behavior as early as possible and adapt the assistance system accordingly.

On the contrary to intra-personalization, inter-personalization is referred to as the differences between drivers. Two drivers can have different driving behaviors given a same situation. With the assumption that individual information about drivers is stored in each maneuver execution. The question being if it is possible to differentiate the drivers by looking at their past maneuver executions.

1.2 Structure of the book

An overview of this book is depicted in Figure 1.2 with the arrows showing suggested paths for reading. Overall, the book are divided into four main parts. In the first part (Part I), the motivations and goals of this book are presented. Along with that, four research questions are addressed.

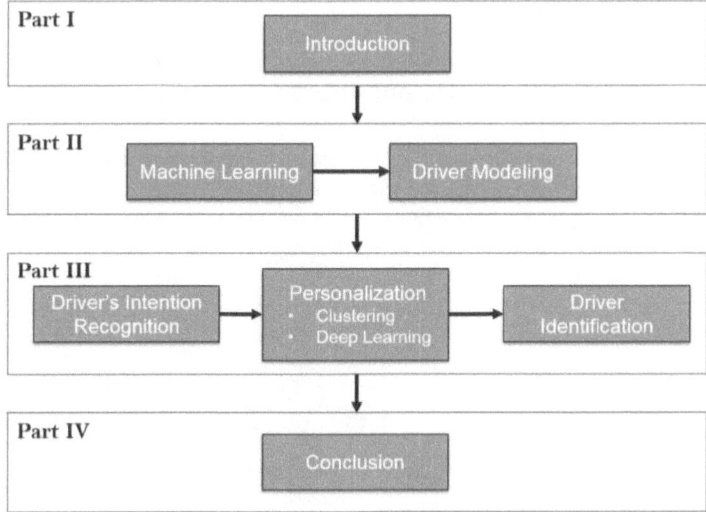

Figure 1.2: Overview of the book structure. Each The arrows connecting parts and chapters show the suggested path for reading the book

The Part II of the book contains two chapters which provide the fundamental background for understanding the book. In particular, the concept of machine learning and the fundamentals of related machine learning methods are then introduced in Chapter 2. Further, since this book also focuses on the personalization in automotive application, Chapter 3 gives an overview of driver modeling and personalization problem in advance driver assistance systems.

The third part (Part III) is the main contribution of this book. It contains three chapters, in which the four research questions presented in Chapter 1 are analyzed and answered in different a spects. C hapter 4 presents the first m ain c ontribution o f t his t hesis i n r ecognizing driver's intention. In this chapter, various machine learning models are compared in the task of prediction lane change intention. Also the first approach on personalization of driver model is analyzed and discussed.

In Chapter 5, a novel approach is presented, that extracts the driver information from maneuver executions and eventually exploits the dependency between maneuver executions to personalize the prediction of driver's preferences when turning left. in the course of this chapter, the answers to the research question 2 and 3 are provided

Chapter 6 dives into the problem of inter-personalization and provide an answer to the fourth research question. In this chapter, we reformulate the problem of driver classification as a comparison problem. Our models based on Siamese architecture are able to extract driver behaviors and identify unseen drivers.

The last part (i. e. Chapter 7) concludes the book and summarizes the answers for four research questions. Possible directions for future works are also discussed in this chapter.

2 Machine Learning

Artificial Intelligence (AI) is one of the newest research fields in computer science. The term artificial intelligence was coined since 1940s. It refers to intelligent agents which are able to make decisions or/and react either humanly or rationally (Russell and Norvig, 1995). The definition of AI is not restricted on how the agent is implemented, whether it is programmed based on a set of well-predefined rules, complex searching algorithms or if it learns from the environment.

Machine learning is considered as a subset of AI, in which the intelligent agent learns to perform a task from experiences. According to Mitchell (1997), a computer program, which is considered as *learning* to solve a task T with regards to a performance metric P, if it is able to improve its performance given the experience E. This is similar to the learning behaviors observed in animal and human being. In machine learning, the experience E is usually represented as a set of recorded data and could be in various formats like numerical vector, time series, images, videos, etc.

The development of machine learning fits well with the tremendously increasing amount of data that human is able to record. Nowadays, we can collect data about customer's behaviors of online shops and websites, data for monitoring vehicles or other sensor-rich devices. This large volume of data exceeds the capability of human to manually process. Machine learning provides a tool to facilitate the automation of the data processing.

There are various machine learning models that aim to tackle diverse learning problems. Those models can be categorized based on the type of feedback that contains in the data (or in the provided experience E) (Russell and Norvig, 1995):

- **Unsupervised Learning** specifies the learning problem when there is no explicit feedback provided. In this case the unsupervised learning algorithms aim to discover the patterns and hidden structure from the data.

- **Supervised Learning** is a set of learning problems in which the learner is given the desired output that it has to predict. The learner observes a set of training samples which consists of the input $\mathbf{x}$ and the output or label y. It will have to learn a function f, that maps the input to the output: $y = f(\mathbf{x})$.

Depending on the value of the intermediate feedback y, supervised learning can be further divided into subcategories: When the value of y is a limited set of classes, learning f is referred to as a classification problem. While continuous value of y results in regression problems.

- In **Reinforcement Learning**, the learner also receives feedback but it is in form of a reward signal. It differs from supervised learning since the reward signal is not the desired output that a learner has to predict. The feedback (reward signal) in this case is the guidance for the learner to discover the optimal actions (Sutton, Richard S. and Barto, 1998).

- **Semi-supervised Learning** is a combination of unsupervised and supervised learning. It refers to a set of learning problems where the labels are only available for a small part of the training data (Russell and Norvig, 1995).

2.1 Unsupervised Learning

In this section, the fundamentals of two main techniques in unsupervised learning are presented: clustering and dimensionality reduction. These two techniques are used in the book for analyzing and visualizing driver's data.

2.1.1 Clustering

Clustering is one of the most common tasks in unsupervised learning (Nilsson, 1997). The goals of clustering algorithms are to discover the hidden structure within the data and assign each data point to a group. The assignments are done in a way such that similar data points are put together in a same group and dissimilar data points are separated into different groups. For example, given a set of unlabeled training instances (gray data points shown Figure 2.1a). A distance-based clustering algorithm can group instances that are relatively close to each others into a same cluster. The clustering algorithm can discover and form two groups of data points which are shown in Figure 2.1b.

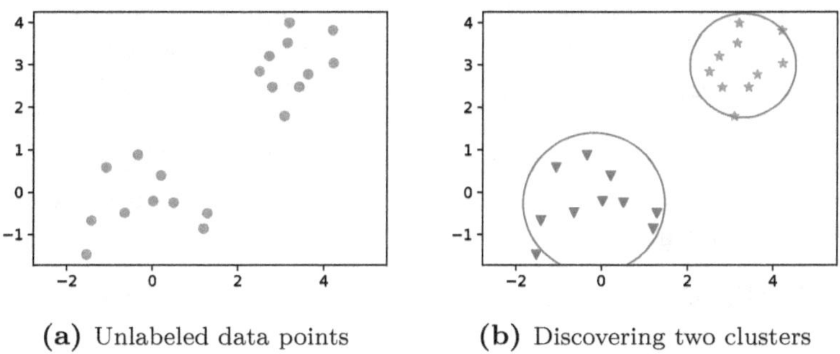

(a) Unlabeled data points   (b) Discovering two clusters

Figure 2.1: Illustration of clustering algorithms

Two main techniques for estimating the similarity between data points are: distance-based and density-based methods. Based on the data and the domain, one may choose to use a specific similarity metrics.

K-means Clustering

K-means is a distance-based clustering algorithm. The objective of K-means is to divide n data points into these k groups such that the sum of the variances of all groups is minimized.

Formally, given a data set of n data points:

$$\mathbf{X} = \{\mathbf{x}_0, \mathbf{x}_1, ..., \mathbf{x}_n\} \tag{2.1}$$

and k is the desired number of disjunctive clusters.

$$\mathbf{C} = \{C_1, C_2, ..., C_k\} \text{ with } C_i \cap C_j = \varnothing, \forall i \neq j \tag{2.2}$$

Furthermore, let μ_i be the mean of all data points that are assigned to the i-th cluster C_i, the objective of K-means is to find the optimal assignment f:

$$f : \mathbf{X} \to \mathbf{C} \tag{2.3}$$

such that the following term is minimized:

$$\sum_{i=1}^{k} \sum_{\mathbf{x} \in C_i} \|\mathbf{x} - \mu_i\|^2 \tag{2.4}$$

Finding Cluster Center: Finding optimal cluster assignment is a NP-hard problem (Flach, 2012). However an approximate solution can be obtained using heuristic algorithms. The well-known algorithm for estimating optimal cluster centers is introduced by (Lloyd, 1982) (This algorithm is also called Voronoi iteration or K-means algorithm). The algorithm initialize the clusters randomly and then iteratively update the cluster assignments. It contains three steps:

- **Initialization:** randomly initializes k cluster centers $\{\mu_1, \mu_2, \ldots, \mu_k\}$.

- **Quantization:** for each data point $\mathbf{x}_j$, the quantization step computes the distance $d_{i,j}$ between μ_i to $\mathbf{x}_j$. Each data point is then assigned to the closest cluster. The assignment function f can be written as:

$$f(\mathbf{x}_j) = \underset{C_i}{\operatorname{argmin}} \, d(\mu_i, \mathbf{x}_j) \tag{2.5}$$

- **Estimation:** this step estimates the new optimal centers given the assignment from the previous step. The new center of the cluster C_i is simply the mean of all data points that belong to this cluster.

$$\hat{\mu}_i \leftarrow \frac{1}{|C_i|} \sum_{\mathbf{x}_j \in C_i} \mathbf{x}_j \tag{2.6}$$

Where $|C_i|$ denotes the number of elements in C_i.

Step 2 and 3 are repeated until the convergence conditions are met or when reaching the maximal number of iterations. Convergence conditions are when the assignment does not change after an iteration or when the cluster centers (μ) only slightly moved their previous positions:

$$|\hat{\mu}_i - \mu_i| < \epsilon, \, \forall i = 1..k \tag{2.7}$$

Initialization Problem: In practice, the presented K-means algorithm converge quickly to a solution, however it cannot guarantee to find a global optimum that minimizes the objective shown in Equation (2.4). One of the problem being bad initialization of the k cluster centers (Selim and Ismail, 1984, Ward Jr, 1963).

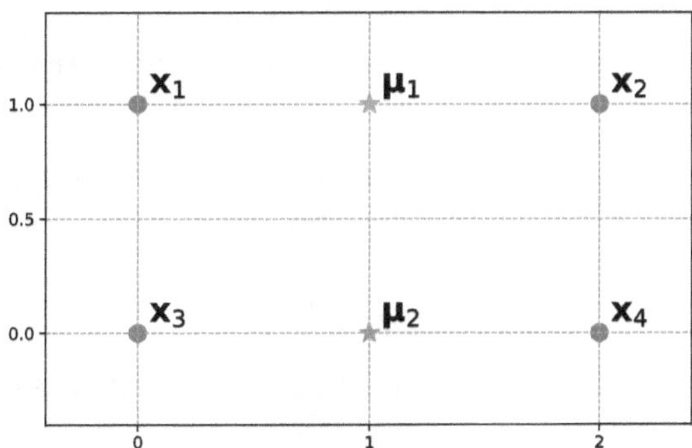

Figure 2.2: An example of bad initialization of the cluster centers (μ_1, μ_2). The optimal solution will not be found in this case.

For example, given 4 data points in a 2-dimensional feature space: $\mathbf{x}_1, \ldots, \mathbf{x}_4$ with $\mathbf{x}_i \in \mathbb{R}^2$; as shown in Figure 2.2. If we apply K-means algorithm with $k = 2$ on this data set and initialize the two cluster centers μ_1, μ_2 as shown in red and green dots, the algorithm will converge immediately and result the following assignments: $C_1 = \{\mathbf{x}_1, \mathbf{x}_2\}$ and $C_2 = \{\mathbf{x}_3, \mathbf{x}_4\}$. This is not the optimal solution with regard to K-means objective. The optimal assignments are $C_1 = \{\mathbf{x}_1, \mathbf{x}_3\}$ and $C_2 = \{\mathbf{x}_2, \mathbf{x}_4\}$ as showing in the right side of Figure 2.2.

To overcome the initialization problem of K-means, several approaches are introduced. A simple approach is to repeat the K-means algorithm multiple times with different initial centers. The final result is the cluster assignment received from the best run (i.e. the Equation (2.4) is minimized). Since the runs are independent from each other, they can be parallelized to improve the speed of the algorithm.

Instead of randomly initializing the cluster centers, one can select the initial centers so that they are well-spread over the data set. To this end, (Arthur and Vassilvitskii, 2007) proposed an algorithm called k-means++, that selects the initial centers in a probabilistic fashion to improve the convergence speed and accuracy in the mean of finding the global optimum. k-means++ selects the initial centers iteratively:

- It randomly choose one of the data points as the first center μ_1.

- The i-th center is chosen from the remaining data points such that the probability of being chosen increases with the distance to the closest center.

Experimental results show that `k-means++` can substantially improve the convergence speed K-means (Arthur and Vassilvitskii, 2007).

2.1.2 Dimensionality Reduction

The term dimensionality in this chapter is referred to as the number of features contained in each data point. In other words, it is the dimension of the feature space. In complex problems, it is often the case that the collected data points are high dimensional. For example, in image analysis or time series, each instance can contain hundreds or thousands or even millions of features. Highly integrated systems like autonomous cars or satellite can generate data in even higher dimensionality since it is a combination of multiple sensors (e. g. vehicle dynamic sensors, cameras, radars, lidars, etc.).

The volume of possible data points increases exponentially as the number of dimensions increase. This problem is referred to as the curse of dimensionality (Bellman, 1966). In addition, the high dimensionality makes difficult to visualize the data set. However, it is often the case that not all the dimensions are equally important or contain the same amount of information.

Dimensionality reduction is a set of methods that aims to reduce the number of dimensions by either removing uninformative features or by replacing the original features with fewer and abstract ones(Van Der Maaten et al., 2009). The dimensions can be extremely reduced to two or three, which facilitates the visualization of the data set in a comprehensible 2D/3D plots. Formally, let $\mathbf{X}$ be the set of original data points:

$$\mathbf{X} = \{\mathbf{x}_0, \mathbf{x}_1, \ldots \mathbf{x}_n\} \text{ with } \mathbf{x}_i \in \mathbb{R}^d \tag{2.8}$$

We want to map each data point of $\mathbf{X}$ on a lower dimensional space $\mathbb{R}^k$, k is much smaller than d. Let $\mathbf{Y}$ be counterpart of $\mathbf{X}$ on $\mathbb{R}^k$:

$$\mathbf{Y} = \{\mathbf{y}_0, \mathbf{y}_1, \ldots, \mathbf{y}_n\} \text{ with } \mathbf{y}_i \in \mathbb{R}^k \tag{2.9}$$

Thereby, $\mathbf{Y}$ should preserve some desired properties of the original data $\mathbf{X}$ such as the distance between data points. By stressing on which properties should be kept, different techniques are developed. Principal components analysis (PCA), for example, is a linear dimensionality reduction technique that preserves the global structure of the data by maximizing the variance of $\mathbf{Y}$. Another approach is to retain the local structure (neighborhoods) as in Stochastic Neighbor Embedding (SNE).

Linear Transformation

A widely used technique in dimensionality reduction is Principal Component Analysis (PCA) introduced by Pearson (1901). This approach generates new features by orthogonally projecting the data onto a lower dimensional space. For example if we have two data points in a 3D-space, the best 1D linear space (a straight line) that retains the distance between two data points is the line that connects them. With the observation that a lower dimensional space can best represent the original data when the variance of the data is preserved, the objective of PCA for finding a new lower dimensional space is to maximize the variance of the projected data.

Formally, let $\overline{\mathbf{x}}$ be the mean of the original data set $\mathbf{X}$ and and $\overline{\mathbf{y}}$ be the mean of the projection $\mathbf{Y}$. Since $\mathbf{Y}$ is a linear projection of $\mathbf{X}$, there is a direction vector $\mathbf{u}$ of $\mathbf{Y}$. $\mathbf{y}_i$ is the projection of $\mathbf{x}_i$ on u and is defined by:

$$\mathbf{y}_i = (\mathbf{x}_i \cdot \mathbf{u})\mathbf{u}, \text{ with } i = 1 \ldots n \text{ and } \|\mathbf{u}\| = 1 \tag{2.10}$$

Let y_i denote the length of $\mathbf{y}_i$.

$$y_i = \|\mathbf{y}_i\| \tag{2.11}$$
$$= (\mathbf{x}_i \cdot \mathbf{u})\|\mathbf{u}\| \tag{2.12}$$
$$= \mathbf{x}_i \cdot \mathbf{u} \text{ (since } \|\mathbf{u}\| = 1) \tag{2.13}$$

The objective of PCA can be then formulated as finding a direction vector $\mathbf{u}$ such that $var(y)$ is maximal:

$$\mathbf{u}_1 = \underset{\mathbf{u}}{\operatorname{argmax}}(var(y)) \tag{2.14}$$

$$= \underset{\mathbf{u}}{\operatorname{argmax}} \frac{1}{n} \sum_{i=1}^{n} (y_i - \overline{y})^2 \tag{2.15}$$

$$= \underset{\mathbf{u}}{\operatorname{argmax}} \frac{1}{n} \sum_{i=1}^{n} \mathbf{u}^T (\mathbf{x}_i - \overline{\mathbf{x}})(\mathbf{x}_i - \overline{\mathbf{x}})^T \mathbf{u} \tag{2.16}$$

$$= \underset{\mathbf{u}}{\operatorname{argmax}} \mathbf{u}^T \Sigma \mathbf{u} \tag{2.17}$$

and Σ is the covariance matrix of the original data set:

$$\Sigma = \frac{1}{n} \sum_{i=1}^{n} (\mathbf{x}_i - \overline{\mathbf{x}})(\mathbf{x}_i - \overline{\mathbf{x}})^T \tag{2.18}$$

Since the direction vector $\mathbf{u}$ is a unit vector, it has the magnitude of 1 ($\|\mathbf{u}\| = 1$). Therefore, using Lagrange multiplier, we can rewrite the optimization problem to the problem of maximizing:

$$\mathbf{L}(\mathbf{u}) = \mathbf{u}^T \Sigma \mathbf{u} - \lambda(\mathbf{u}^T \mathbf{u} - 1) \tag{2.19}$$

$\mathbf{L}(\mathbf{u})$ reaches its maximum when the partial derivative of with respect to $\mathbf{u}$ is 0:

$$\frac{\partial \mathbf{L}}{\partial \mathbf{u}} = 0 \tag{2.20}$$

$$\rightarrow \Sigma \mathbf{u} - \lambda \mathbf{u} = 0 \tag{2.21}$$

This is exactly the definition of eigenvector ($\mathbf{u}$) and eigenvalue (λ). In additional, when the Equation (2.21) is met, the corresponding eigenvalue λ is the variance of the projected data on the eigenvector $\mathbf{u}$. The problem of finding the principal components is becoming finding the eigenvector and eigenvalue of the covariance matrix Σ. The eigenvector corresponding to the largest eigenvalue is the first principal component.

Figure 2.3 shows a set of 2D data points together with two principal components (PC1 and PC2). The length of each principal vector is variance of the data points projected on that respective dimension.

Stochastic Neighbor Embedding (SNE)

In comparison to PCA, SNE is a non-linear dimensionality reduction method that aims to preserve local structure of the data. When using lower dimensional space to represent a higher one, we are losing some amount of information. It means not every relationship between data points can be preserved. SNE weights the relationship between close neighbor points more than between two data points that are far away.

The first version of stochastic neighbor embedding (SNE) was introduced by (Hinton and Roweis, 2002). To illustrate the idea of SNE, let consider a data point $\mathbf{x}_i$ of $\mathbf{X}$ and its counterpart $\mathbf{y}_i$ of $\mathbf{Y}$. SNE assumes that the probability of a data point $\mathbf{x}_j$ being picked as neighbor of $\mathbf{x}_i$ is a Gaussian distribution, whose center lies at $\mathbf{x}_i$ and has a standard deviation of σ_i. Based on this assumption, the conditional probability that $\mathbf{x}_j$ being picked as neighbor of $\mathbf{x}_i$ is defined as:

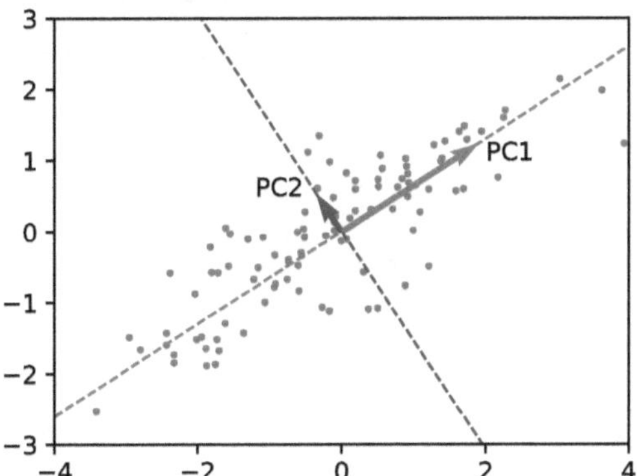

Figure 2.3: Two principal components found in a 2-dimensional data set. The variance of the data is maximized along the first principal component (PC1)

$$p_{j|i} = \frac{\exp(-d_{ij}^2/2\sigma_i^2)}{\sum_{k\neq i}\exp(-d_{i,k}^2/2\sigma_i^2)} \tag{2.22}$$

with d_{ij} is the euclidean distance between x_i and x_j:

$$d_{ij}^2 = \|\mathbf{x}_i - \mathbf{x}_j\|^2 \tag{2.23}$$

Note that the denominator of Equation (2.22) normalizes the probability so that $\sum_j p_{j|i} = 1$. With that, we have defined the probability distribution over the neighbors of each data point $\mathbf{x}_i$:

$$P_i = \{p_{j|i}\}, \text{ for } j = 1\ldots n \tag{2.24}$$

Likewise, SNE assumes that there is a Gaussian distribution centered at each data point y_i in the low dimensional space with standard deviation of $\frac{1}{\sqrt{2}}$. Note that the standard deviation is fixed in the low-dimensional space. The probability of picking a data point y_j as neighbor of y_i is defined as:

$$q_{j|i} = \frac{\exp(-\|\mathbf{y}_i - \mathbf{y}_j\|^2)}{\sum_{k\neq i}\exp(-\|\mathbf{y}_i - \mathbf{y}_k\|^2)} \tag{2.25}$$

This in turn defines the probability distribution over the neighbors of y_i in the low-dimensional space:

$$Q_i = \{q_{j|i}\}, \text{ for } j = 1\ldots n \tag{2.26}$$

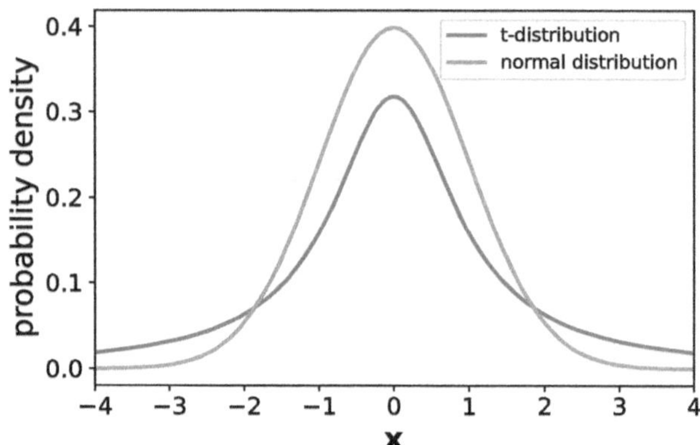

Figure 2.4: The probability density function of the t-distribution with 1 degree of freedom and of the normal distribution

SNE's Objective: Since $\mathbf{X}$ is given, the probability P_i of each data point $\mathbf{x}_i$ is fixed and can be computed using Equation (2.22). The objective of SNE is now to find an optimal placement of $\mathbf{Y}$ in the low dimensional space $\mathbb{R}^k$ such that Q_i is close or equal to P_i.

To optimize the placement of $\mathbf{Y}$, SNE uses Kullback-Leibler divergence (Kullback, 1997) to measure the difference between P_i and Q_i: $KL(P_i||Q_i)$. The objective of SNE thus becomes to minimize $KL(P_i||Q_i)$ for every all data point i. This can be formulated as a total Kullbeck-Leibler divergences of all data points:

$$C = KL(P||Q) = \sum_i KL(P_i|Q_i)$$
$$= \sum_i \sum_j p_{j|i} log \frac{p_{j|i}}{q_{j|i}}$$

(2.27)

One property of using Kullbeck-Leibler divergence is that modeling a large $p_{j|i}$ by deploying a small $q_{j|i}$ can aggravate the results more than modeling a large $q_{j|i}$ by using a large $p_{j|i}$. That means breaking the local structure around a data point is more costly than breaking the global structure. In other words, wrong placement of data points that are far is neglected.

The loss function C is minimized using gradient descent. That is, $\mathbf{Y}$ is firstly positioned randomly on the low-dimensional space. The loss function

C and its gradient are then computed at each data point y_i:

$$\frac{\partial C}{\partial \mathbf{y}_i} = 2\sum_j (\mathbf{y}_i - \mathbf{y}_j)(p_{i|j} - q_{i|j} + p_{j|i} - q_{j|i}) \qquad (2.28)$$

The gradient of C shows the direction to which $\mathbf{y}_i$ should be adjusted so that C is minimized. Let call $\mathbf{y}_i^{(t)}$ is the value of $\mathbf{y}_i$ at the t-th iteration.

$$\mathbf{y}_i^{(t)} = \mathbf{y}_i^{(t-1)} + \lambda \frac{\partial C}{\partial \mathbf{y}_i} \qquad (2.29)$$

Improving SNE using t-Distribution: t-Distributed stochastic neighbor Embedding (t-SNE) was introduced by (Maaten and Hinton, 2008). t-SNE propose two changes to alleviate crowding problems and also reduce the complexity of the optimizing problem. Instead of trying to modeling the conditional probability of a point j given i (i.e. $p_{j|i}$) in the first place, it optimizes the placement of $\mathbf{Y}$ to keep the joint probability p_{ij} of original data points. This joint probability is defined as:

$$p_{ij} = \frac{p_{j|i} + p_{i|j}}{2n} \qquad (2.30)$$

with $p_{j|i}$ is given in Equation (2.22). That means P and Q are now the joint probability distribution over neighbors in the original space $\mathbf{X}$ and the low dimensional space $\mathbf{Y}$. In comparison to the conditional probability, the joint probability is symmetric (i.e. $p_{ij} = p_{ji}$). This probability facilitates the calculation of the gradient of the loss function.

$$\frac{\partial C}{\partial \mathbf{y}_i} = 4\sum_j (\mathbf{y}_i - \mathbf{y}_j)(p_{ij} - q_{ij}) \qquad (2.31)$$

The next modification in t-SNE aims to alleviate the crowding problem. Instead of using Gaussian distribution to compute the probability of neighborhood in the lower dimensional space, t-SNE deploys the student's t-distribution with 1 degree of freedom. The reason being, t-distribution with 1 degree of freedom is a heavy tail distribution (cf. Figure 2.4). This property allows data points to spread out in the low dimensional space. Moreover, the computation of q_{ij} using t-distribution is faster since it does not involve the exponential term as in Gaussian distribution. Formally, q_{ij} is defined as:

$$q_{ij} = \frac{(1 + \|\mathbf{y}_i - \mathbf{y}_j\|^2)^{-1}}{\sum_{k \neq l}(1 + \|\mathbf{y}_k - \mathbf{y}_l\|^2)^{-1}} \qquad (2.32)$$

A notable disadvantage of SNE and also in t-SNE is that this method is non-parametric. Once an optimal placement of $\mathbf{Y}$ is found, it cannot map a new incoming data point into the lower dimensional space without having to recompute the whole placement of $\mathbf{Y}$. In this book, t-SNE is therefore only used for visualizing a given data set.

2.2 Supervised Learning

In supervised learning, in addition to the input $\mathbf{x}$, the algorithm also receives the target variable y which it has to predict (Russell and Norvig, 1995). For example, $\mathbf{x}$ is a picture and y is whether there is a cat in the picture. The objective of supervised learning is then to find/learn a function f that can map the input to the output:

$$y = f(\mathbf{x}) \tag{2.33}$$

There are various ways to approximate the function f. Each method has its strengths and weaknesses. Depend on the nature of the problem one may choose a specific method over the others. This chapter gives an overview of methods to estimate f, that are used in the scope of this book.

2.2.1 Support Vector Machine

Consider a binary classification problem with the training data as shown in Figure 2.5a. Assume that we are solving this problem using a linear model. That is, we are trying to find a possible line that can separate the class A from class B. In this case, there are multiple solutions that can satisfy the separation requirement. However not all the models are equally good. The idea of support vector machine is to determine the best solution among all possible ones. To do this, it uses the concept of the margin to define the goodness of a model. The margin of a model is the shortest distance between the separation line and data points from training data. The larger the margin is, the better the model becomes (Boser et al., 1992, Cortes and Vapnik, 1995).

Formally given a data set with n training samples: $(\mathbf{x}_i, y_i)$, $i = 1 \dots n$ with $\mathbf{x}_i \in \mathbb{R}^m$ is the m-dimensional feature vector and $y_i \in \{-1, 1\}$ is the label. A linear hyperplane in the feature space can be written as:

$$\mathbf{w} \cdot \mathbf{x} - b = 0, \text{ with } \mathbf{w} \in \mathbb{R}^m \tag{2.34}$$

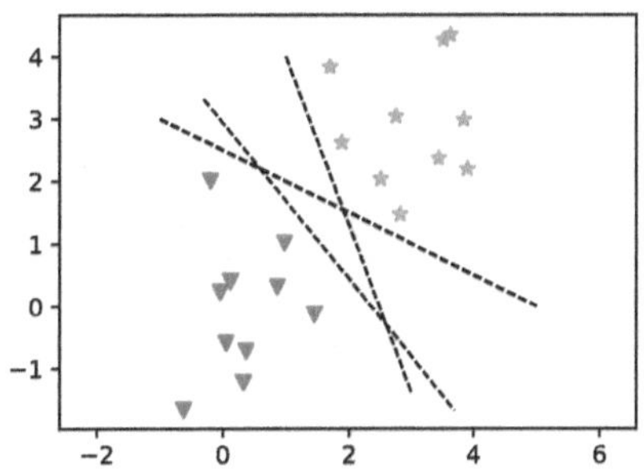

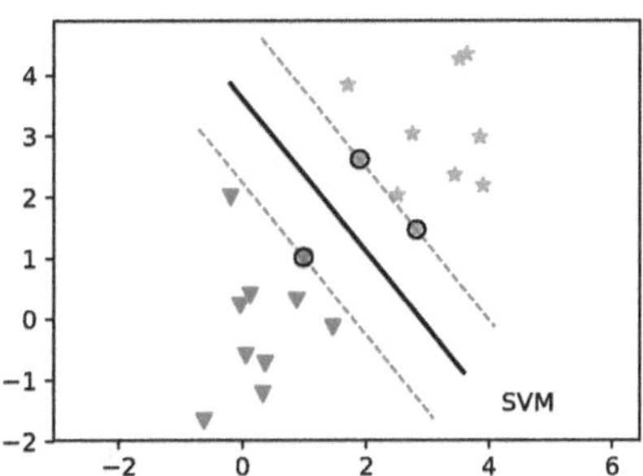

(a) Multiple solutions for separating two classes

(b) SVM maximizes the margin between two green lines

Figure 2.5: A support vector machine finds optimal hyperplane for separating two classes by maximizing the distance between nearest sample and the hyperplane.

Assume that the data is linearly separable and there are multiple hyperplanes that separate the data into classes, the problem of maximizing the margin (i. e. finding the optimal solution) can be written as:

- Find two parallel hyperplanes: $\mathbf{w} \cdot \mathbf{x} - b = 1$ and $\mathbf{w} \cdot \mathbf{x} - b = -1$, that can separate the data points into two desired classes

- Maximize the distance between these two hyperplanes, which is $\frac{2}{\|w\|}$. Maximizing this term is equivalent to minimizing $\|w\|$.

Without loss of generality, we can assume that all data points that belong to the class $y_i = 1$ lie above the first hyperplane and data point of class $y_i = -1$ lies under the second hyperplane. These two conditions can be formulated as:

$$\begin{aligned} \mathbf{w} \cdot \mathbf{x}_i - b \geq 1 \quad \text{when } y_i = 1 \\ \mathbf{w} \cdot \mathbf{x}_i - b \leq -1 \text{ when } y_i = -1 \end{aligned} \tag{2.35}$$

Which can be summarized to a single condition:

$$y_i(\mathbf{w} \cdot \mathbf{x}_i - b) \geq 1 \text{ for } i = 1 \ldots n \tag{2.36}$$

The problem now become a constrained optimization problem in which we want to minimize $\|\mathbf{w}\|$ subject to the constraint (2.36). This can be formulated as a Lagrangian function.

$$L(\mathbf{w}, b, \lambda) = \frac{1}{2}\|\mathbf{w}\|^2 - \sum_{i=1}^{n} \lambda_i(y_i(\mathbf{w} \cdot x + b) - 1) \tag{2.37}$$

with $\lambda_i \geq 0$. The problem of finding the optimal hyperplane now become maximizing $L(\mathbf{w}, b, \lambda)$ with respect to λ and minimizing L with respect to $\mathbf{w}$ and b:

$$min_{\mathbf{w},b}\, max_\lambda L(\mathbf{w}, b, \lambda) \tag{2.38}$$

We can switch to solve the dual problem of Equation (2.38) since it delivers the same result. The dual problem is formulated identical as in Equation (2.38) but *min* and *max* terms are switched.

L is minimized with respect to $\mathbf{w}$ and b when the its partial derivatives is 0. Solving $\frac{\partial L}{\partial \mathbf{w}} = 0$ and $\frac{\partial L}{\partial b} = 0$ results in:

$$\mathbf{w}^* = \sum_{i=1}^{n} \lambda_i y_i \mathbf{x}_i \tag{2.39}$$

$$0 = \sum_{i=1}^{n} \lambda_i y_i \tag{2.40}$$

Note that $\mathbf{w}$ still depends on λ which can be found by maximizing L. Use Equation (2.39) and (2.40) to shorten (2.37) results in:

$$L(\lambda) = \frac{1}{2} \sum_{i=1}^{n} \sum_{j=1}^{n} \lambda_i \lambda_j y_i y_j \mathbf{x}_i \mathbf{x}_j + \sum_{i=1}^{n} \lambda_i \tag{2.41}$$

The objective is now to find λ that maximizing L subject to the constraints: $\lambda_i \geq 0$ and $\sum_{i=1}^{n} \lambda_i y_i = 0$. This is again a quadratic problem and can be solved using sequential minimal optimization (Platt, 1998). Switching from the primal problem (i.e solving Equation (2.37))to the dual one (Equation (2.41)) does not necessary make it easier, but it has a nice property that allows us to apply the kernel tricks.

Kernel Trick (Guyon et al., 1993) Up to this point, the described SVM is used to find a linear hyperplane to separate the data points. The kernel trick allow us to efficiently convert the original feature space of $\mathbf{x}$ to a higher dimensional $\phi(\mathbf{x})$. The product $\mathbf{x}_i\mathbf{x}_j$ in Equation (2.41) now becomes $K(\mathbf{x}_i, \mathbf{x}_j) = \phi(\mathbf{x}_i)^T\phi(\mathbf{x}_j)$. Note that computing the kernel

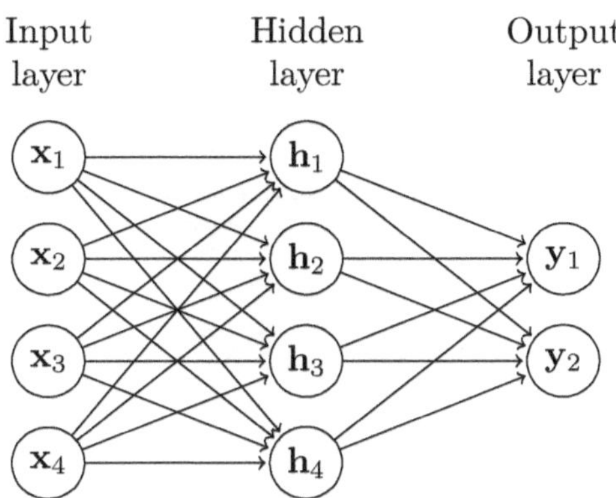

Figure 2.6: A feed-forward neural network with one hidden layer

function K is much more efficient than compute the transformation $\mathbf{x} \to \phi(\mathbf{x})$. For example in the case of radial basis function kernel (rbf):

$$K_{\mathbf{rbf}}(\mathbf{x}_i, \mathbf{x}_j) = \exp(-\frac{\|\mathbf{x}_i - \mathbf{x}_j\|^2}{2\sigma^2}) \tag{2.42}$$

$K_{\mathbf{rbf}}$ can be directly computed efficiently but its corresponding transformation function $\phi_{\mathbf{rbf}}(\mathbf{x})$ has infinite dimension.

2.2.2 Artificial Neural Networks (ANN)

Artificial Neural Networks were proposed since 1943 by McCulloch and Pitts (1943). ANNs are inspired by the human's brain, it mimics the way information is transferred between biological neurons. An ANN uses artificial neurons to simulate the biological neurons. Multiple artificial neurons are connected together to build up an ANN. It can be viewed as a directed graph in which the artificial neurons are connected by a weighted edge. The weights represent the importance of information that being transferred from one artificial neuron to another. An artificial neuron can receive information from multiple nodes. The inputs are then transformed within the receiving node by an activation function. This function decides how many information is being forward to the next node. An artificial neuron can be viewed as a function $\delta(\cdot)$ of n input nodes $\{\mathbf{x}_1, \mathbf{x}_2, ..., \mathbf{x}_n\}$:

$$y = \delta(w_1 \cdot \mathbf{x}_1 + w_2 \cdot \mathbf{x}_2 + ... + w_n \cdot \mathbf{x}_n + b) \tag{2.43}$$

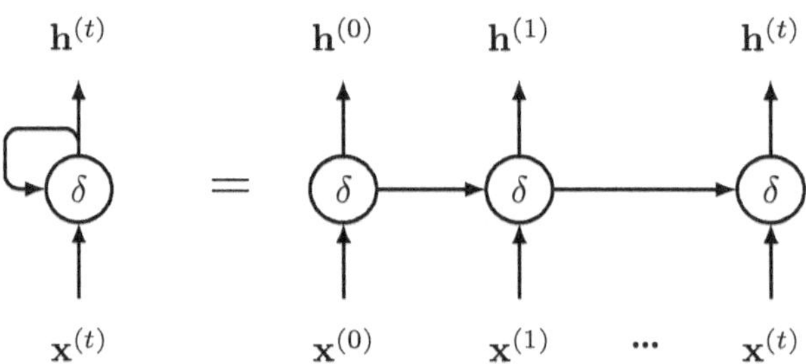

Figure 2.7: Visualization of the recurrent connection in a RNN. The right side is a unrolled version of the left side.

with w_i ($i = 1 \ldots n$) is the weight of each connection and b is the bias. w_i and b are the learnable parameters of an ANN.

The artificial neurons can be stacked up that creating a complicated ANN which is difficult to train. An popular approach to simplify the ANN architecture is to divide artificial neurons into groups (so called layers). Information is then transformed and transferred between the layers. By stacking the layers on top of each others, it forms a feed-forward neural network (Figure 2.6). The most-left layer is the input layer, the most-right layer is the output layer that produces the final prediction of the network. Layers that lie between the input and output layers are called hidden layer.

Recurrent Neural Network

Recurrent neural network (RNN) is a special type of ANN. It is designed to process sequential data (Hopfield, 1982, Rumelhart et al., 1986). The input of RNN is not just the current sample of the data but also the results from processing previous time steps. To capture the dependency between time steps, RNN introduces a loop in the network. That is, the output from previous time step is used as additional input when processing the next time step. Let variable t denote the time step counter ($t \in \mathbb{Z}$). Each data point is now associated with a time step and will be denoted as $\mathbf{x}^{(t)}$. Furthermore, let $\mathbf{h}^{(t-1)}$ be the hidden state from the previous time step, RNN can be defined using following recursive formula:

$$\mathbf{h}^{(t)} = \delta_h(W \cdot \mathbf{x}^{(t)} + U \cdot \mathbf{h}^{(t-1)} + b) \tag{2.44}$$

$$o^{(t)} = \delta_o(T \cdot \mathbf{h}^{(t)} + c) \tag{2.45}$$

with W and U are the weight matrices that are used to compute the hidden state at time step t from input of the current time step ($x^{(t)}$) and hidden state of previous time step ($\mathbf{h}^{(t-1)}$) respectively. b is the bias for computing $\mathbf{h}^{(t)}$. $o^{(t)}$ is the (intermediate) output at time step t, which is computed using the weight matrix T and the bias c. $\delta_{\mathbf{h}}$ and δ_o are the activation functions, for example we can use $tanh(.)$ for $\delta_{\mathbf{h}}$ and $softmax(.)$ for δ_o when learn a classification task. In case, the intermediate outputs are not importance then the Equation (2.45) is only applied on last time step.

A traditional RNN contains 3 weight matrices $\{W, U, T\}$ and two bias vectors $\{b, c\}$. Because of the loop connection between consecutive hidden states, it can be unroll to form a very deep network (see right side of Figure 2.8). That can cause a problem when training the RNN using back-propagation since that loss from the last time step n must be propagated back to the first time step. With a large n, it causes the gradient to be vanished or exploded. To better depict this problem, we can rewrite the Equation (2.44) as:

$$\mathbf{h}^{(t)} = F(U \cdot \mathbf{h}^{(t-1)}) \tag{2.46}$$

This is a multiplicative Let's assume that we using back-propagation to optimize the weight matrix W. The network make some errors at the last time step t. Formally, to optimize parameters $\theta = \{W, U, b\}$ of the network using back-propagation, we need to compute the derivative of loss with respect to θ. This involves the computation of the partial derivative of $\mathbf{h}^{(t)}$ over θ.

$$\frac{\partial \mathbf{h}^{(t)}}{\partial U} = \frac{\partial \mathbf{h}^{(t)}}{\partial \mathbf{h}^{(t-1)}} \cdots \frac{\partial \mathbf{h}_1}{\partial \mathbf{h}_0} \frac{\partial \mathbf{h}_0}{\partial U} \tag{2.47}$$

$$= \left(\prod_{i=t\ldots 2} \frac{\partial \mathbf{h}^{(i)}}{\partial \mathbf{h}^{(i-1)}} \right) \frac{\partial \mathbf{h}_1}{\partial U} \tag{2.48}$$

with

$$\frac{\partial \mathbf{h}^{(i)}}{\partial \mathbf{h}^{(i-1)}} = U \cdot F'(U \cdot \mathbf{h}^{(i-1)}) \tag{2.49}$$

With that, the Equation (2.48) contains U^{t-2} in it. When $\|U\| > 1$, the gradient explodes with the increasing number of time steps t. And likewise, if $\|U\| < 1$ the gradient vanishes when t is large. In other words, the network is not able to capture the long term dependency between time steps.

Long Short-Term Memory

Training a recurrent neural network (RNN) on data with a long time lag is difficult. The reason being that the gradient could either explode or vanish when it is back-propagated through time. Long short-term memory (LSTM) is an extended RNN architecture that was introduced by (Hochreiter and Schmidhuber, 1997) to address this problem. LSTM overcomes the gradient vanishing and exploding by using memory cells together with gating mechanisms. The memory cell is a vector $(c^{(t)})$ whose purpose is to store information from previous time steps. Which information being stored in the memory is regulated by the gating mechanism.

Gating Mechanism: The gate is a mechanism to manipulate the flow of information. A gate is an activation function (δ), which takes as input the feature vector of current observations $(\mathbf{x}^{(t)})$, the hidden state $(\mathbf{h}^{(t-1)})$ from the last time step. The most popular activation used in a gate is sigmoid function since its output range lies between 0 and 1.

$$\text{sigmoid}(\mathbf{x}) = \frac{1}{1 + e^{-\mathbf{x}}} \tag{2.50}$$

A value of 0 means that the gate completely blocks the information flow and a value of 1 lets all of the information go though. There are three types of gates used in the LSTM architecture:

- **Input Gate** learns to weight the new memory proposal of the current time step:

$$\delta_i^{(t)} = \delta(W_i \mathbf{x}^{(t)} + U_i \mathbf{h}^{(t-1)} + b_i) \tag{2.51}$$

- **Output Gate** learns to weight the output before forwarding it to the next timestep:

$$\delta_o^{(t)} = \delta(W_o \mathbf{x}^{(t)} + U_o \mathbf{h}^{(t-1)} + b_o) \tag{2.52}$$

- **Forget Gate:** $\delta_f^{(t)}$ learns to discard a memory cell from previous time step. This allows outdated information to be deleted from the memory.

$$\delta_f^{(t)} = \delta(W_f \mathbf{x}^{(t)} + U_f \mathbf{h}^{(t-1)} + b_f) \tag{2.53}$$

The respective weight matrices $W_{\{i,o,f\}}$, $U_{\{i,o,f\}}$ and the bias vectors $b_{\{i,o,f\}}$ are learnable parameters, which will be optimized during the learning process.

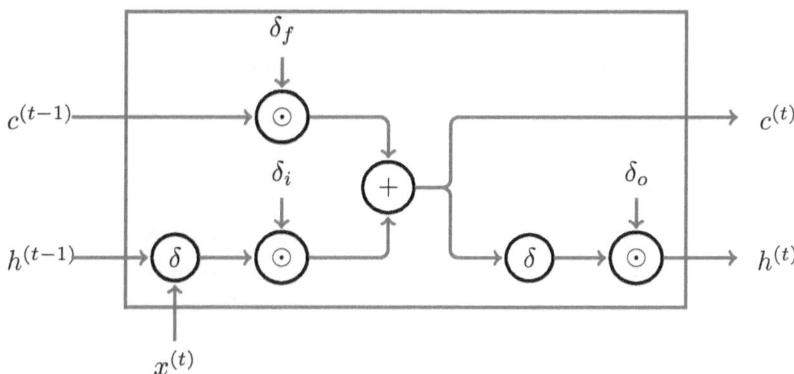

Figure 2.8: Visualization of information flow in a long short-term memory cell at time step t. $\odot$ is the element-wise multiplication of two inputs. $+$ is the element-wise addition. δ is the non-linear activation function.

Memory Cell: At each time step t, the network compute the potential contribution to the memory cell from the inputs (the current input value from time step t and the hidden state from time step $t - 1$):

$$\tilde{c}^{(t)} = \delta(W_c\mathbf{x}^{(t)} + U_c\mathbf{h}^{(t-1)} + b_c) \tag{2.54}$$

Note that, in this case, δ is not a gating function and can be any activation function. Mostly, the hyperbolic tangent function (tanh) is used to compute the memory cell proposal $\tilde{c}^{(t)}$:

$$\tanh(\mathbf{x}) = \frac{e^{\mathbf{x}} - e^{-\mathbf{x}}}{e^{\mathbf{x}} + e^{-\mathbf{x}}} \tag{2.55}$$

The new memory cell $c^{(t)}$ is then computed as a sum of the memory cell of last time step and the contribution from the input at current time step:

$$c^{(t)} = \delta_f^{(t)} \odot c^{(t-1)} + \delta_i^{(t)} \odot \tilde{c}^{(t)} \tag{2.56}$$

Here, the forget gate $\delta_f^{(t)}$ regulates how strong the memory cell from the last time step ($c^{(t-1)}$) is being forward to $c^{(t)}$. In addition, the input gate $\delta_i^{(t)}$ regulates how much the inputs at time step t contributes to $c^{(t)}$. The gate is applied on information flow by using element-wise multiplication $\odot$. The Equation (2.56) is the key concept of LSTM for dealing with gradient vanishing and exploding problem. Instead of using multiplication to model recurrent of information as in RNN, LSTM allows the memory from the previous time step to be added directly to the current memory. This can be clearly observed when comparing the RNN recurrent equation (2.48) with the LSTM memory update equation (2.56).

Output at Each Time Step: At each time step, the intermediate output of the LSTM network is computed from its memory cells by using an activation function (δ) and is regulated by the output gate $\delta_o^{(t)}$:

$$h^{(t)} = \delta_o^{(t)} \odot \delta(c^{(t)}) \tag{2.57}$$

The activation function δ is usually the hyperbolic tangent function (tanh).

Figure 2.8 illustrates how the components of LSTM are connected together and how the information flows inside a LSTM cell. To simplify the graphic, the inputs of all three gates described in the equations (2.51), (2.52) and (2.53) are not shown. As we can also observe, the path connecting $c^{(t-1)}$ and $c^{(t)}$ is short and regulated by the forget gate. This connection forms a shortcut and allows information to be stored and transferred between time steps.

With these extensions, the LSTM network architecture is able to learn to remember the relevant information of time points in the past and use them for later predictions. Recently, LSTM networks have achieved some state-of-the-art results on time series data (Sutskever et al., 2014), acoustic modeling (Beaufays et al., 2014) and machine translation problems (Hirschmann et al., 2016).

2.2.3 Random Forest

Random forest is a well-known implementation of ensemble method (Breiman, 2001, Ho, 1995). In general, the idea of ensemble methods is to learn multiple weaker models (base models) instead of trying to learn one single powerful model. The prediction of an ensemble is then the combination of predictions from all models. Ensemble methods make use of two assumptions:

- The performance of base model is better than random (e. g. in term of accuracy).

- The base models are independent. That means, the predictions of one model are not correlated with the predictions of other models.

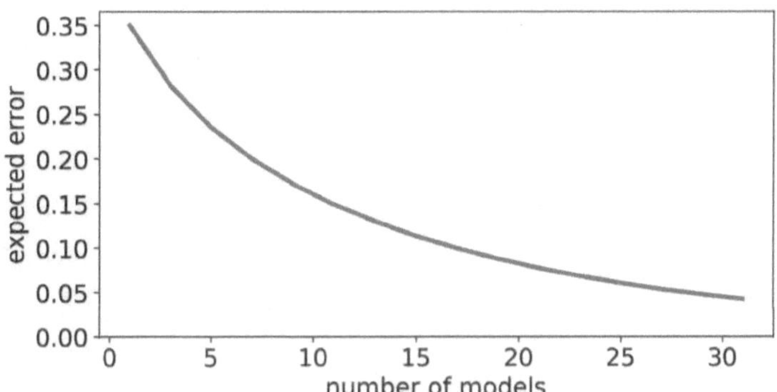

Figure 2.9: Expected error of ensemble model with respect to the number of models in the ensemble

It can be proved that if these two assumptions are met, the prediction performance of an ensemble will increase proportionally to the number of base models. In fact, given an ensemble containing $2n + 1$ base models and the error rate of each model is ϵ ($\epsilon < 0.5$). The ensemble emits a wrong prediction when the majority (i. e. $n + 1$) of the base model produce a wrong prediction. With the independence assumption, this probability can be computed as:

$$p(n, \epsilon) = \sum_{i=n+1}^{2n+1} \binom{2n + 1}{i} \epsilon^i (1 - \epsilon)^{2n+1-i} \tag{2.58}$$

whereas $\binom{2n+1}{i}$ is the number of i-combinations from $2n + 1$ possible models. This term is computed as:

$$\binom{2n + 1}{i} = \frac{(2n + 1)!}{i! \, (2n + 1 - i)!} \tag{2.59}$$

For example, when building an ensemble from 25 base models and each base model has an error rate of 0.35, the ensemble's error rate $p(n, \epsilon)$ reduce to 0.06 which is much smaller than $\epsilon = 0.35$. Figure 2.9 shows the expected error rate of an ensemble over the number of models.

Bootstrap Aggregating (Bagging)

In practice, the base models are not truly independent since they are learn from a same data set. Breiman (1996) introduced Bagging as a technique

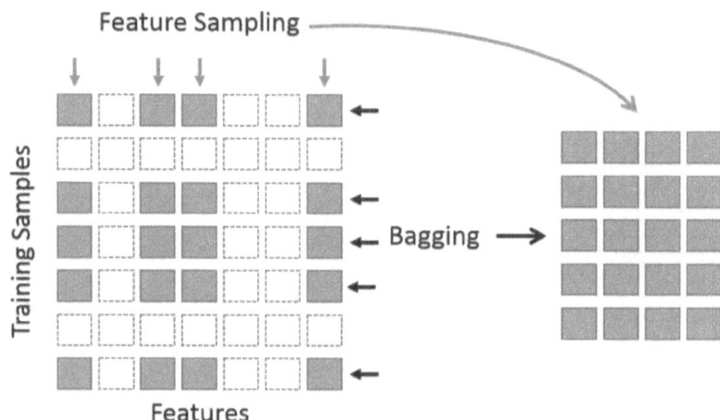

Figure 2.10: Visualization of bootstrap aggregating (bagging) and feature sampling for selecting a subset from the data set. The original training data set is visualized on the left. Each row represents a training sample and each column a feature. On the right side, the selected data set after applying bagging and feature sampling

that can be applied to train less-dependent base models. The idea is not to train each base model on the whole data set but only on a subset of the data. To draw a subset from the original data set, bagging uses bootstrap samples (i. e. random sampling the data set with replacement).

Bootstraps samples allows it to have duplicates when sampling, that is, one sample can be selected multiple times or missed when constructing a subset of samples. In fact, if n is the size of the original data set, the probability of one particular sample not being selected when exactly n samples are picked is:

$$p(n) = (1 - \frac{1}{n})^n \tag{2.60}$$

We also have $\lim_{n \to \infty} p(n) \approx 0.368$, which means about 36.8% of the original data set will not appear in the sample. This property makes the subsets become diverse. Models trained on these subsets are become less-dependent. The independent assumption is therefore strengthen which enhances the performance of the final ensemble prediction.

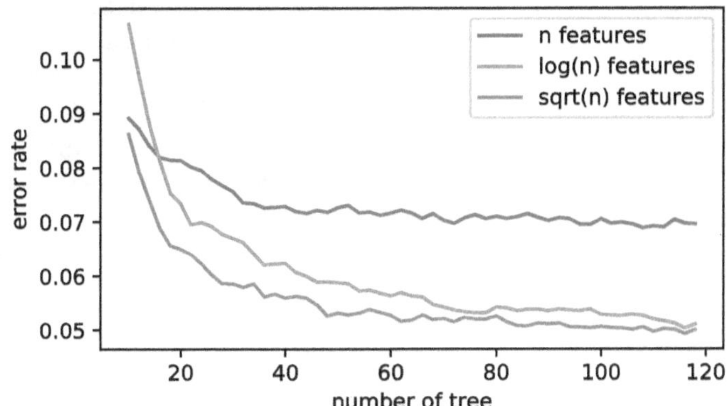

Figure 2.11: Comparison between different feature sampling strategies with n is the total number of the features.

Feature Sampling

To further improve the diversity in base models, random forests use another technique called feature sampling (or subspace sampling). This approach randomly chooses a subset of the features to train the each base model (Ho, 1998). To compare bagging and feature sampling, we can view the data set as a table in which each row is a training sample and each column is a feature. Bagging technique will randomly select the rows and feature sampling will randomly select the columns of the table (see Figure 2.10)

The base model used in random forest is decision tree. While training a random forest, an alternative implementation of feature sampling is usually used. Instead of sampling a subset of feature before training a each decision tree, it applies features sampling when looking for best split. That is, for each node of the decision tree, a subset of features is drawn. The condition for splitting is then only searched in this subset of the feature set. This approach can further increase the diversity of the single model (i. e. decision tree) and thus improves the prediction of the resulting ensemble.

Figure 2.11 shows the effect of different feature sampling strategies in building a random forest classifier. The experiment is conducted on MNIST[1] data set with the number of features is equal the number of pixel

(i. e. $n = 28 \times 28 = 784$). The x-axis show the number of tree in the ensemble. The y-axis show the error rate evaluation on the test set. Using a subset of features to find optimal split shows a clear improvement over using all features. The ensemble benefits from the increasing number of trees when using feature sampling in comparison to using all features. In addition, constructing a tree using feature sampling is much faster than when using all features.

3 Driver Behavior Modeling and Personalization

Personalization is one of the key to increase the user experience and service quality. It has been addressed in various contexts, including online marketing, recommender systems and driver assistance systems. This chapter firstly presents an overview of the personalization problem in general. Subsequently, different approaches of personalization in the domain of driver assistance systems are discussed. Finally, the driver assistance system developed in project PRORETA 4 is introduced, which aims at providing personalized recommendation for drivers.

3.1 Personalization

The term personalization is not a new concept, it refers to the process of *tailoring a product or service* so that they meet the *needs of a particular person*. Considering this definition, there are two important aspects on which we should pay attention to: "tailoring a product or service" and "the needs of a particular person". The former indicates that products and services have to possess a certain level of flexibility so that they can be adjusted to the user's needs. The latter refers to the recognizing of customers' needs.

The flexibility of physical products is limited, which leads to production on demand or mass production with different variations of a same product (e. g. in clothes or shoes industry). Softwares or digital services, on the other hand, are much more flexible, they can be customized with little effort. Examples for such products are customizable websites that allow users to change the theme or layout; or news websites in which users can specify topics they are interest in (such as My Yahoo![1] or Google News[2]). Personalization has been an important feature of major websites and web services. In the domain of automotive application, personalization is also considered as a key to improve user experience (Hasenjager and Wersing, 2017, Ponomarev and Chernysheva, 2019, Sarkar and Kortge, 2006).

Going beyond the ability to let users specify and customize the product by themselves, personalized systems also aim at automatically detecting and inferring user's needs and preferences. This kind of advanced systems can further improve the user experience and comfort. Commercially, such personalized systems are already available in the domain of web search and online marketing (Arora et al., 2008, Hannak et al., 2013). In web search, the position of each URL on the search engine result pages is ranked according to the prediction whether the user would be interested in that URL. In online marketing, the recommendations of shopping items are derived from the click and purchase history of the user. Users receive personalized recommendations without having to explicitly specify their preferences.

In all of the mentioned applications, the behavioral data is an essential part for personalized systems since it reveals users and their preferences. Nowadays, behavioral data can be recorded from users at high speed, in large volume and from different sources. To cope with the enormous amount of data and gain knowledge from it, machine learning is appeal to be a suitable tool. With the help of machine learning algorithms, complex behavioral patterns in the data can be detected and future behaviors can be predicted.

Detecting drivers' preferences has been recently a focus in the domain of automotive applications. This information is of great benefit to the assistance systems in supporting drivers at their driving tasks. The ability to understand the drivers and to provide them with personal recommendations will not only improve the comfort but also the safety.

3.2 Driver Behavior Modeling

As it has been shown, human errors are the cause for most of traffic accidents. A better understanding of driver behavior plays therefore an important role in increasing traffic safety. However, driver behavior modeling is a complex and broad topic since participating in traffic includes different levels of human behaviors. It involves from high level strategy like planing the route to microscopic level such as small adjustments of the steering wheel to keep the vehicle remaining in a desired lane.

In this section, we will firstly review the qualitative categorizations of driving behaviors and the various factors that influence the driver behav-

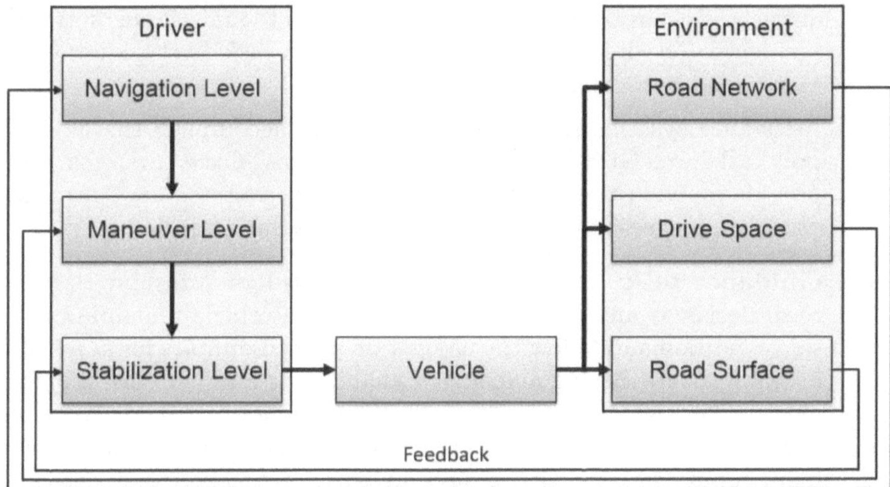

Figure 3.1: Three level hierarchical classification of primary driving tasks according to Donges (1982).

ior. This promotes a better understanding of drivers and the their relations with the ego-vehicle and the environment. Furthermore, this systematic categorization of driver behaviors is also the foundation to identify potential applications of driver assistance systems. Subsequently an overview of two sub-problems in driver behavior modeling is presented, namely: driver intention prediction and driving style recognition.

3.2.1　Categorization of Driver Behaviors

There are different researches that aim at systematically categorize the driver behaviors from either psychological or physiological point of views. A well-known approach introduced by (Donges, 1982) divides the driver's activities into three hierarchical levels: *Navigation, Guidance* and *Stabilization*. These three levels relate to the time horizons from the moment when the decision of executing a task is made to the point when the task is accomplished.

- **Navigation task**: This level describes long-term tasks that involve planning and choosing a suitable route. The navigation task can be done before the driver physically participates in the traffic, but it can also occur during a trip. An example for the later case is when the driver decides to change the route because of traffic jams or traffic accidents. The information requires for this task is the knowledge about road network and traffic information.

Assistance systems that support drivers to accomplish the navigation tasks are GPS-based navigation systems. Other systems that provide mobile communications such as radios, traffic news are also considered as assistance systems for the navigation tasks.

- **Guidance task**: Activities of drivers at this level are short to mid-term decisions and executions that keep the vehicle remaining in a safe condition and follow the navigation plan. Examples for activities at this level are driving maneuvers: changing to another lane, keeping a safety distance to the leading vehicle, turning maneuvers, etc. The perception of current traffic situation surrounding the vehicle is the main input that the driver uses at this level. The exact decision (e. g. when and how to execute a lane change maneuver) depends additionally on the individual driver's traits.

  Nowadays, there are already numerous assistance systems available in commercial vehicles supporting driver at this level. For examples: adaptive cruise control, lane departure warning, blind spot warning, etc.

- **Stabilization task**: At this level, the driver performs microscopic and small adjustments that keep the vehicle stable and follow the decision that has been made at guidance level. Inputs for this level are for example the properties of road surface or the deviation between the current trajectory and the desired one. Systems that support drivers at stabilization task are for example anti-lock braking system (ABS) or electronic stability control (ESC).

These three levels are hierarchical and built on top of each other. The outputs of the current level are treated as the inputs of the succeeding one. The actual interactions between driver and the vehicle are realized at the lowest level (the stabilization level).

A similar but more granular categorization can be found in (Hamdar, 2012). Here, driver tasks are broken down into five levels, in which, navigation tasks are subdivided into *Pre-trip* and *Strategic En Route*. Guidance level is split into *Tactical Route Execution* and *Operational Driving*. Finally *Vehicle Control* level can be mapped to the stabilization level as in (Donges, 1982).

It is pointed out by both authors, the guidance level (which is equivalent to tactical route execution and operational driving levels in (Hamdar, 2012)) is the most promising level for the development of advanced driver assistance systems (Donges, 2016, Hamdar, 2012). Maneuvers executed by drivers at this level intermediately influence the road traffic and thus directly affect the safety of drivers.

3.2.2 Driver's Intention Prediction

As mentioned in Chapter 1, human factor is found as the major cause of traffic accidents. A straight forward approach to reducing traffic accidents is to anticipate the next actions that drivers would perform. Base on that, driver assistance systems can analyze potential risks of that situation and warn drivers in advance or even intervene in the maneuver executions if necessary. In this line of thought, we focus on the driver's intentions on the maneuver level. These are intentions of performing maneuvers such as changing lane, turning, stopping etc. Being able to predict such maneuvers several seconds before they happen will be of great benefit to the traffic safety.

How a driver assistance system can benefit from driver's intention prediction? Let's review blind spot assist, which is already introduced in production vehicles[3]. The blind spot assist system makes use of side radars to detect other vehicles that are in the blind spot of the ego-vehicle. Depending on whether the driver is about to perform a lane change or turn maneuver or not, the system can respectively rise a suitable warning. For example when there is another vehicle in the blind spot but the driver doesn't show any intention to change lane, then the system will just show a small LED signal to let driver know about the object in the blind spot. When the driver confirms the direction indicator to change lane, the system will rise a more critical warning with acoustic signal to notify driver of the danger.

In the described system, the driver explicitly expresses her intention through the indicator signal. However, a study conducted by Olsen (2003) shows that there are about 66 % of drivers use the direction indicator when changing lane. That means relying on this signal will fail to detect the other 34 % of total number of lane change cases. Even in the case that drivers activate the direction indicator, 50 % of the times the indicator is activated

---

[3]The name "Blind Spot Assist" is used by Daimler AG, other similar systems can be found in the market like "Blind Spot Detection", "Blind Spot Warning" or "Blind Spot Information System" etc.

only shortly before or even during the lane change maneuver (Lee et al., 2004). A more sophisticated method to deal with this problem is to incorporate a lane markings detection system. It provides information on the relative position of ego vehicle and the lane markings, and thereby enables the prediction of potential changing lane without using indicator signals. To this end, information about vehicle dynamics of ego vehicle are also employed as inputs. Signals such as longitudinal, latitudinal acceleration, velocity and yaw angle are also used for this purpose.

To further improve the prediction of driver's intention, more sensors can be deployed to directly observe the driver. For example, using interior cameras to capture the gaze/head direction. This additional information can help the system better recognize the driver's intention and provide a trustworthy prediction on time. In fact, in Chapter 4, we will shows that it is possible to not just predict the lane change intention of drivers but also to determine when the driver will perform this several seconds beforehand.

The advancement of sensing technology provides essential tools to monitor the driving tasks and also the drivers. However, the increasing number of sensors also poses a challenge to build more powerful predictive models that are capable of processing and utilizing the vast amount of information. Machine learning has been seen as a superior approach for this kind of tasks as it is mostly applied regarding the prediction of driver intentions.

3.2.3 Driving Style Recognition

A widespread approach on personalizing driver assistance systems is to recognize the driving style and adjust the systems accordingly. The information about driver type or driving style is valuable to optimize the vehicle for different purposes such as reducing fuel consumption, recognizing driver's aggressiveness or improving traffic safety (Martinez et al., 2018).

Driving style recognition simulates how humans would normally evaluate and assign driver behaviors into different categories (e. g. aggressive, normal and calm). Since driver behavior is an umbrella term, the definition of driving styles is also very abstract. There are several definitions of driving styles that were introduced in (Elander et al., 1993, Sagberg, 1999, Warner et al., 2011). They all share the same point of view that driving styles are the individual driving habits that a driver chooses to drive and that are developed over time.

This definitions cover all possible behaviors of driver and thus are qualitative, they do not explicitly specify how to evaluate and measure driving styles or how many groups of driving styles are there. There is a need for frameworks that can quantify the driving styles so that it can be used to further improve driver assistance systems.

To this end, there are two main approaches that are usually employed to identify driving styles in literature: subjective and objective (data driven) approaches. The subjective approaches are conducted by analyzing the self-assessment from drivers. On the other hand, objective approaches rely mostly on data generated by driver while driving.

Subjective Approaches

To evaluate the driving styles, subjective approaches utilize self-assessments. Typically, the self-assessment is designed in form of a questionnaire. The answers for the question are in the form of multiple choices options. The possible answers are either yes/no or a scale that the driver can choose from (e. g. from "strongly agree" to "strongly disagree" or from "very frequently" to "never").

French et al. (1993) developed a driving style questionnaire to assess driver behaviors on six different dimensions:

- Speed: if the driver drives faster than the speed limit on motorway or on built up areas.

- Calmness: where or not the driver remains calm or reacts to other drivers.

- Social Resistance: assess the behavior of driver when receiving advice from others.

- Focus: assess the ability to deal with distraction (e. g. performing secondary tasks while driving).

- Planning: if the driver plans the trip ahead (i. e. routing or places for taking break). This actually corresponds to the driver behavior at navigation level as mentioned in Section 3.2.1.

- Deviance: e. g. whether the driver overtakes on the wrong side or skips the red light.

By analyzing the correlation between each dimension with the accident rates, it shows that speed has the highest influence on the accident rate, especially for driver who is younger than 60 year old. "Planning" has on the other hand positive impacts on reducing the accident rate, this means drivers who usually plan there trip and intermediate stops ahead have a lower accident rate. This kind of approach, however, suffers from the subjective perceptions of driver about her own driving styles. For example, the perception of aggressiveness and calmness may vary from driver to driver.

König et al. (2002) summarized the results of project S.A.N.T.O.S, in which several approaches for identifying driver types are presented. The five driving types are derived from the literature and are analyzed on test drivers:

- sporty/aggressive

- dynamic/progressive

- experienced/calm

- conservative/relaxed

- anxious/defensive

The driving styles of test drivers are determined by using a standardized questionnaire which is filled by a third person (the investigator). It differs from other traditional questionnaire-based approach since all test drivers are rated by the same investigator. This would make the resulting driving styles more comparable.

The results of subjective approaches provide a better understanding of driving styles and can be used as the basic for developing further methods. However this approach is not suitable to be embedded into a final product since it would require each customer to fill up a questionnaire. This also does not cover the fact that driver behaviors can change over time. A driver would have a different driving style when driving to vacation comparing with his everyday commuting.

Data-driven Approaches

Objective approaches are referred to data-driven approaches, in which the driving styles are recognized or discovered by analyzing the driving data. For this purpose, vehicle dynamic signals are mostly used as input for recognizing and classifying driving styles. The inputs are usually extracted from inertial measurement unit (IMU) together with other control signals like steering wheel, gas and brake pedals position. The vehicle dynamics can also be extract from GPS devices, as proposed in (Constantinescu et al., 2010).

Apart from using sensors that are mounted on the vehicle, there are also works that aim at building a standalone software for smartphone platforms. They utilize the IMU of smartphones to collect driving data and classify driving styles (Corti et al., 2013, Fazeen et al., 2012, Johnson and Trivedi, 2011). The advantage of this approach includes the ease of data collection process since it is a standalone platform and there is no modification in the vehicle required.

In contrast to subjective based approaches, data-driven approaches do not require driver to complete the self-assessment. Instead, the driving styles are implicitly derived from the data. This allows to build applications that are capable of recognizing driving style on the fly, which is more suitable as a product for end users.

The driving data are usually broken down into driving events at maneuver level, which is the second level of the hierarchical categorization introduced in Section 3.2.1. The most used events are turning, roundabout, lane changing and car-following. On top of that, smaller events like braking/decelerating and accelerating are also considered as inputs for recognizing driving styles. Also when conducting a trip, drivers usually have to repeatedly perform different maneuvers at distinct traffic situations. This accelerates the data collection process and makes it possible to observe and compare driver behaviors in various situations.

Methodology for Data-driven Approaches There are different methods for assessing driving style. Most of these method fall back on the knowledge obtained from objective methods. The most straight forward approach is to model the correlation between the driving style and the vehicle dynamics using a set of rules. Murphey et al. (2009) proposed an approach that evaluates the ratio of mean jerk of the current driver over the expected mean jerk of the current road segment. Two thresholds are empirically determined to classify driving styles into 3 possible classes: calm, normal and aggressive. This approach is found to work effectively for identify the driving styles with regard to the fuel consumption.

Driving style recognition can be formulated as a classification problem. For this, data sets with ground-truth are needed. (Karginova et al., 2012) collected such a data set from five bus drivers with either normal or hard (more aggressive) driving styles. Among four tested algorithm: decision tree, random forest, k-nearest neighbor and neural network, neural network outperform others in term of accuracy. Note that, in this work, the models are and evaluation process are performed on each time step, the temporal information is ignored. The accuracy score can be further improved by 6 % when averaging the predictions over 30-seconds window.

Constantinescu et al. (2010) present a data-driven approach which makes use of hierarchical clustering and principal component analysis (PCA) to discover six different driver groups. This approach does not make assumptions about driving styles but separates the driver groups by exploring and finding the intrinsic structures of the data set. Since this method is unsupervised, the discovered clusters do not have an explicit meaning but one can assign suitable labels to each cluster by looking at the features that describe them.

Rosenfeld et al. (2015) adapt the driver model of (Fancher and Sayer, 1996) to divide drivers into three groups, and use this information together with demographic features for predicting the driver's preferences in the use of adaptive cruise control (ACC). The driver groups in this study are determined by rules that are manually defined based on empirical observations in the data set. It is shown that both driver type and demographics information have positive impacts on model and enhance the prediction accuracy. This shows the potential of personalization for predicting gap acceptance and for driver modeling in general.

Correlation with Subjective Approaches The categorizing driver into different group is still an important approach that follows both subjective and data-driven approaches. We can also see that many data-driven approaches are originated from or partly use the results of the subjective approaches. In fact, analyses on the results of both type of approaches shows that there are correlations between vehicle dynamics and the assignment of driving style using questionnaire.

By analyzing the vehicle dynamics and the driving style classification, (Marstaller et al., 2002) showed that there is the positive correlation between the maximal acceleration and the aggressiveness of driving style. The classification of driving style was obtained by using the self-assessment questionnaire introduced by (Assmann, 1985).

The study conducted in (Ishibashi et al., 2007) found that there is a significant correlation between the classification obtained by driving style questionnaire (DSQ) and the recorded driving data. The correlation of driving style are found not only with acceleration but also with other control signals like brake and gas pedal position, and with the car following distance. Similar results on correlation between recorded driving data and results of self-assessments are reported by West et al. (1993) and Farah et al. (2009).

3.3 PRORETA 4: Safety by Learning

3.3.1 A Brief History of PRORETA

PRORETA[4] is a long-term research program between Technische Universität Darmstadt and Continental AG. The program is broken down into multiple successive projects. After the final demonstration of project PRORETA4 in October 2019, the project PRORETA5 is now running. With the ultimate goal of making road traffic safer and more comfortable, PRORETA projects investigate in new approaches and technologies for developing and improving driver assistance systems. Hereby, different aspects of driver assistance systems are studied: perception, planing, human machine interface, machine learning, system integration etc.

The first PRORETA project (2002-2006) focused on developing an active assistance system that automatically detects static as well as moving objects and determines if they are obstacles. Dangerous situations caused by obstacles can therefore be identified. When driver does not react to the danger properly, the developed system can intervene and perform an emergency braking maneuver, or if necessary an obstacle avoidance maneuver (Bender et al., 2007a,b, Darms and Winner, 2005).

The second project of PRORETA took place from 2006 to 2009 with the focus on assisting drivers in overtaking maneuvers on two-way country roads. Based on front-cameras and radars, the developed system can observe the position of leading and oncoming objects. In combination with the vehicle dynamic information of ego car, an impending maneuver can be evaluated if it is safe. Upon detecting risky overtaking maneuvers, the system can then either warn the driver or actively intervene the maneuver execution. (Isermann et al., 2016, 2012)

The third project - PRORETA 3 (2011-2014) - introduced two concepts, namely: the *Safety Corridor* and the *Cooperative Automation*. The concept of *Safety Corridor* is derived from the environment information. The

Figure 3.2: User interface for left turn application. Gaps are visualized in the digital instrument cluster. The color of each gap shows the corresponding recommendation of the system for the current driver. Red indicates a too small gap, whereas suitable gaps are colored in green.[5]

system will monitor the driving tasks performed by driver in the background. When the vehicle is about to leave this safety corridor, the system will raise a warning and possibly intervene in the driving tasks. The concept of *Cooperative Automation* allows the system to take over the vehicle stabilization tasks. Drivers can cooperate with the system and therefore only need to perform the driving task on higher levels (Bauer et al., 2012, Winner et al., 2016).

3.3.2 Overview of PRORETA 4

PRORETA 4 is the fourth project, which aims at leveraging the machine learning methods to further increase the safety and comfort of driver assistance systems. Going beyond the traditional approaches of driver assistance systems, whose functionalities cannot be changed once delivered to customers, the goal of PRORETA 4 is to develop a disruptive ADAS. The new ADAS should be able to quickly and automatically adapt itself to the driver. Hereby, the system should learn from the driver behaviors to provide the suitable recommendations. Meaning, along with the environment perception, that has been in the center of interest in the previous projects, the driver perception and driver behaviors also play an importance role in PRORETA 4.

---

[5]This figure is taken from the press release of PRORETA 4 final live demonstration

Use Cases

The use cases of PRORETA 4 focus on supporting drivers in urban situations where the ego-vehicle does not have the right of way. In such situation, drivers have to wait for until the traffic is clear or there is enough space to perform the intended maneuver. The statistics of traffic accidents in Germany reported by Statistisches Bundesamt (2018) shows that over 68 % of road traffic collisions with personal injury occur in urban area. This accounts for more than50 % of severely injuries and more than 30 % number of casualties. Most of the accidents in urban area are related to intersection situations. Two most accident-prone situations are related to intersections.

The main use case of PRORETA 4 is left turn situations where drivers have no right of way and have to yield to oncoming traffic. In this scenario, drivers must wait for a suitable gap that is big enough so that she can safely perform the intended left turn maneuver. The assistance system developed in PRORETA 4 automatically recognizes such situations and provides drivers with recommendations. The recommendations are suggestions of suitable gap, which are visualized using on the cluster display in form of dynamically flowing bars. The color of the bars switches between red and green, depending on the safety degree of the gap (cf. Figure 3.2). The second use case concerns traffic at roundabouts. With similar HMI settings, the system can also provide suggestions for entering roundabouts.

Different from the first use case, in the third use case, we consider the situations, in which the driver approach an intersection with left-yields-right. If the driver doesn't comply to the predefined safety conditions of the system, warnings will be fired. Else, when all the conditions are fulfilled, the driver gets compliments from the system.

Research Topics in PRORETA 4

Scheduled from 2015 to 2018, PRORETA 4 is made up with three different institutes of TU Darmstadt. To be able to recognize the traffic situation is the requirement to support the driver. For this purpose, precise localization is the essential information. Over the course of the project, several new methods for camera-based localization and mapping are developed to further improve the localization quality (Boschenriedter et al., 2018, Luthardt et al., 2017, 2018, 2019).

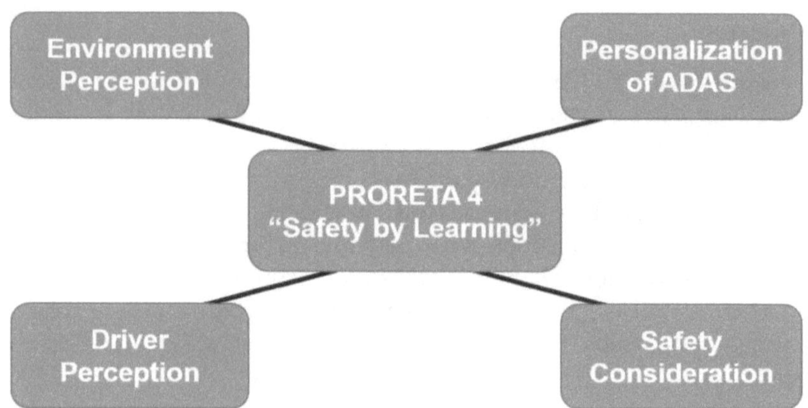

Figure 3.3: Four research topics in PRORETA 4 which serve as building blocks for the final integrated assistance system.

Not only do we consider the surrounding environment in PRORETA 4, but also pay attention to the perception of driver's visual cognition. This delivers another the channel of information that helps the system better understand and support drivers in the decision making process (Schwehr et al., 2019a, Schwehr and Willert, 2017, 2018a,b). In left turn and round-about situations, the system detects whether the driver are distracted by secondary tasks. The driver's distraction can act as a trigger to fire warning and notification messages, when suitable (i. e. green) gaps for turning appear. In addition, the system can also evaluate the securing behaviors of the driver on approaching left-yields-right intersections. The securing behaviors comprise of reducing speed and checking for oncoming vehicles. When any of the securing conditions is violated, the system will raise a corresponding warning.

The personalization of driver models concerns capturing information about the current driver and incorporating it with the environment information to provide personalized recommendations to the driver. For this purpose, various machine learning approaches are used to extract driver information from the data. In this **book**, we will delve deeply into this topic, especially in Chapter 4, Chapter 5 and Chapter 6.

The safety approval of learning system is also of a great concern when applying machine learning methods in the automotive context. To this end, an approach with four step is developed that help to identify lack of generalizability in machine learning models. These four steps cover the whole process of developing a machine learning model: checking data quality, screening for possible violation, evaluating the functional correctness and sensitivity. Researches on this topic can be found in (Henzel, 2019, Henzel et al., 2017).

4 Driver Intention Recognition

One of the keys components to personalize and improve driver assistance systems is to capture the driver behaviors and use that to infers impending actions. Predicting driver's actions in general, and the problem of predicting an impending lane change in particular have been studied in the community under different aspects and set-ups. Mostly, the problem is approached as a classification problem, in which each class represents a possible action in the next few seconds. This chapter re-defines the task as a regression problem, which serves to predict the time until the ego-vehicle touches another driving lane.

Long short-term memory (LSTM) networks are used to capture the past several seconds of driving, extract the driver behaviors and learn to predict the time left until lane change maneuvers. Even though a regression-based formulation of the problem captures more information and should thus be harder to learn, it can be empirically demonstrated that the proposed network provides slightly better results than a comparable classification-based approach. Moreover, the additional information about the exact time of the impending lane change could be used to further increase a user's acceptance of advanced driver assistance systems. In this chapter, the personalization of the network is also conducted by fine tuning it to a particular driver's behavior. The fine-tuning results in an improvement of F1-Score while observing about 20 minutes of driving data from that driver.

4.1 Introduction

One of the most interesting fields in the automotive community is to be able to understand the driver and predict her next actions. The ability to foresee what a driver is going to do could help the system to analyze the potential risk of that situation and warn the driver in advance. Such a system can improve the traffic safety tremendously and avoid many accidents that are caused by human errors.

On top of that, the more information about the possible next action of the driver and when he would perform it, the better we can design and adjust driver assistance systems to not only increase safety but also comfort. A possible application is to adapt the warning time point and warning strategy. This ultimately can help reducing the number unnecessary warnings, thus increase the trust of drivers on the systems.

Changing and merging lanes together with turning are maneuvers at guidance level (cf. Section 3.2). They are among the most common causes for traffic accidents (McGwin, Jr and Brown, 1999), the reason being that drivers overlook other vehicles or do not check the blind spot before performing intended maneuvers.

With current sensor technology, vehicles are able to detect if in the current situation a lane change is safe or not. In some standard car equipment, blind-spot checking systems and lane-keeping systems are already integrated. In order to enable the driver to react to a dangerous situation in time, an advanced driver assistance system (ADAS) should be able to predict the future maneuver in few seconds before it would actually happen. The ADAS can then combine this information with the environment information to determine whether a warning is necessary.

In this chapter, we re-formulate the problem of predicting lane changes as a regression problem, in which the model should learn to predict the actual time-to-lane-change (TTLC). To solve this problem, long short-term memory (LSTM) networks are deployed to capture the information from past time steps. The LSTM architecture are then connected to two deeper layers for predicting TTLC. The developed networks are then validated using a data set recorded using driving simulators. Figure 4.1 shows a situation in which the driver approached a slow leading vehicle (the small white car further ahead on the right lane) and was preparing for a lane change maneuver. The trained LSTM network can anticipate this information and output a prediction for TTLC at 2.7 seconds before the lane change with an error of 0.2 seconds.

The remainder of this chapter is structured as follows. In Section 4.2, we review related approaches for lane change prediction proposed in the literature. In Section 4.3, we formulate the lane-change prediction problem and the TTLC prediction problem. Our proposed approach is then presented in Section 4.4. An evaluation on a data set recorded from 34 test drivers is presented in Section 4.5. Section 4.6 provides the summarization and a conclusion for this work.

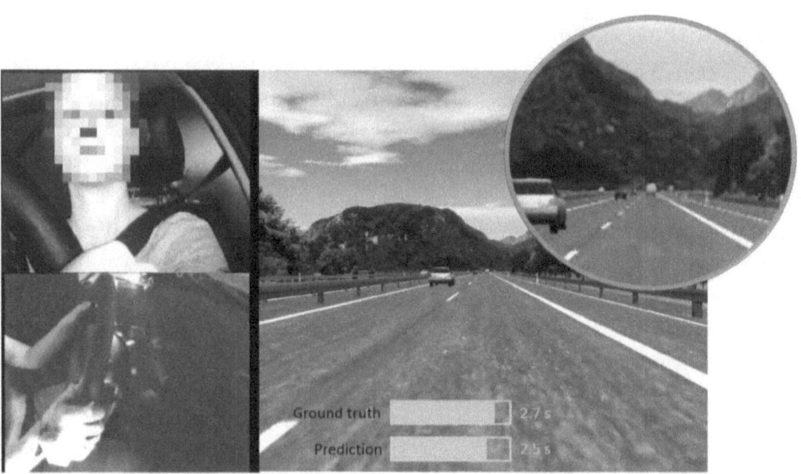

Figure 4.1: Visualization of Time-to-Lane-Change prediction few seconds before an actual lane change. The green bar shows the ground truth TTLC and the orange bar shows the prediction of our regression model.

4.2 Related Work

In the literature, driver intention prediction and maneuver prediction are sometimes a source of confusion. The ground truth for the actual driver *intention* of changing the lane is hard to capture. It could be the case, that a driver wants to change the lane but the traffic situation does not allow the maneuver to be performed. In such a situation, the driver's intention is hard to label without her extra input. In fact, it is only possible to infer driver's intention after the she realized it, but impossible when she was not able to. In this chapter, the driver's intentions are referred to as the realizable intentions of drivers. Eventually, the to-be-realized intentions are more safety relevant to the drivers.

Many approaches have been introduced for modeling the driver behavior. In (Lefèvre et al., 2014), the authors proposed a system using two level of hidden Markov models (HMMs) for predicting unintentional lane departures. At the first level, three separated HMM models for lane keeping as well as left and right departures are learned. At the second level, a merging model is trained to simulate the transitions between the three models at the first level.

Berndt (2016) also proposes to use Hidden-Markov models HMM to recognize driver intention to change lane. The proposed architecture is trained using the whole maneuver executions. By reducing to linear HMMs, the author can decompose the trained model into smaller models. Each of the smaller models is correspond to different stage of a maneuver execution. With that an early recognition of the maneuver start are possible. The evaluation of the introduced model is reported on maneuver-level. Which means a maneuver is considered as correctly recognized if the model omits the right prediction in at least one time step during the whole maneuver execution. This can lead to the fact a model can have high accuracy but the prediction is only produce shortly before (or even after) the ego vehicle touches the lane marking. Interestingly, the results of (Berndt, 2016) shows that their models can predict lane change to the right better than when predicting lane change to the left which is opposite to the result found in this work. However the study is conducted on a different dataset. It is partly recorded in the city traffic and has no information about the gaze and head direction of the driver.

Jain et al. (2015) proposed an extended version of HMMs called Autoregressive Input-Output HMMs, which consist of a hidden layer, an output layer and in addition an input layer. The input layer represents the environmental features, and the output layer additionally incorporates the information from a driver camera that captures the driver's face and head movements. An autoregressive connection in the output layer is introduced to handle the temporal correlation of these movements. When the algorithm predicts a maneuver it will keep this prediction for the next five seconds and evaluate it as a true positive if this maneuver occurs at any time in this five-second period. In this work, we predict the maneuver for every time step (which is about 25 prediction/second) and evaluate the predictions against the ground truth received from the labeling process described in Section 4.4.2.

Jain et al. (2016) introduced a sensor fusion network architecture that is based on recurrent neural networks (RNNs) for anticipating lane-change and turning maneuvers. The proposed architecture uses two different RNNs for processing sensor data from inside and outside features. Several dense layers are then used for fusing the outputs from these two RNNs. In addition, a new loss function is introduced, that also takes into account the number of maneuver observations that has been fed into the network. The idea of this loss function is similar to the effect of the regression approach taken in this chapter, in that it penalizes the learner more for a wrong prediction the closer it is to a lane change. However, in the regression approach, the model could provide additional information about time-to-lane-change for further assistance.

In (Mandalia and Salvucci, 2005) the lane change prediction problem is approached using support vector machines (SVMs). Kumar et al. (2013) proposed an improvement to the SVM approach, in which a Bayesian filter was applied on the probability output from the multi-class SVM. It explicates that the final predictions become smoother and the number of false positive errors (false alarms) decreases. This Bayesian filter approach could also be used to further improve the predictions of the classifier in our approach.

Morris et al. (2011) analyzed the performance of lane change prediction methods in laboratory and on-road settings. A relaxed constraint for prediction was also applied so that a positive prediction lying within ± 1 second around the ground truth is also considered to be correct. The authors also presented a multi-suppression technique for combining consecutive positive predictions. It was reported that this technique can significantly reduce the number of false positive predictions. However, it can be seen as a trade off between early detecting a lane change and false positive errors. This technique is a layer on top of the existing model. When applied, it requires to see the a certain number of positive predictions from the model before finally outputs a positive prediction.

In (McCall et al., 2007), the lane-change-intent prediction problem was approached using a sparse Bayesian learning method, where the used features also originated from three input channels: lane information, driver monitoring and vehicle data. The case, where drivers wanted to change the lane but were not able to, was also addressed and resolved by labeling such situations to lane change events. To this end, two datasets are recorded, one for lane keeping and one for lane changing. In the first data set, the driver should not attempt to make any lane change maneuver whereas the later one only consists of lane changes in the whole recording. Nevertheless, we have to keep in mind that such a data collection process could affect the driver behaviors and the data would not reflex their natural way of their driving behaviors.

Three intensive analyses on predicting lane change maneuvers are presented in (Alekseenko et al., 2019). All three approaches are developed in the course of the first edition of Intelligent Transportation Systems (ITS) - Data Mining Hackathon. By analyzing the effect of input windows sizes on the prediction performance, it is showed that a window size from 3 to 4 seconds has the best trade off between performance and model complexity. All three approaches use decision-tree based ensemble methods for classifying lane changes. Random forest and gradient boosting trees are both heavily employed since they require less training time. Although the potential of recurrent neural networks are recognized, this kind of models require a large amount of computational power and time to train. In this chapter we also compare decision-tree based models with the proposed network and shows that recurrent neural networks can achieve a better performance for predicting lane changes and time-to-lane-change.

4.3 Problem Formulation

The lane change problem could be formulated as: given the data from driver, vehicle and environment till current time point, the model should predict the probability that the driver will change to another lane in next few seconds.

At every time point t, the data that are observed from driver, ego vehicle and environment are stored in a vector $\mathbf{x}^{(t)}$. Given the current time t_0, the data from the last k time steps until t_0 can be then written as:

$$X^{(t_0),k} = (\mathbf{x}^{(t_0-k+1)}, \mathbf{x}^{(t_0-k+2)}, ..., \mathbf{x}^{(t_0)}) \tag{4.1}$$

The *lane-change prediction* problem is to predict whether a lane change will be performed within next m seconds. The learner should learn a function f of $X^{(t_0),k}$ so that:

- $f(X^{(t_0),k}) = left$, if there is a **left** lane change maneuver within the next m seconds

- $f(X^{(t_0),k}) = right$, if there is a **right** lane change maneuver within the next m seconds

- $f(X^{(t_0),k}) = straight$, if there is **no** lane change maneuver within the next m seconds

The labels: *left*, *right* and *straight* are usually encoded using a numerical values such as -1, 1 and 0 respectively.

Here, we are not only interested in the next lane change but also when exactly the driver will perform that. To do that, we have to reformulate this classification formulation to the problem of predicting time to lane change. Instead of learning one function that predicts the lane-change as discrete classes, we can learn for each type of the lane change (*left* and *right*) a specific function of the input $f(X^{(t_0),k})$. These functions predict

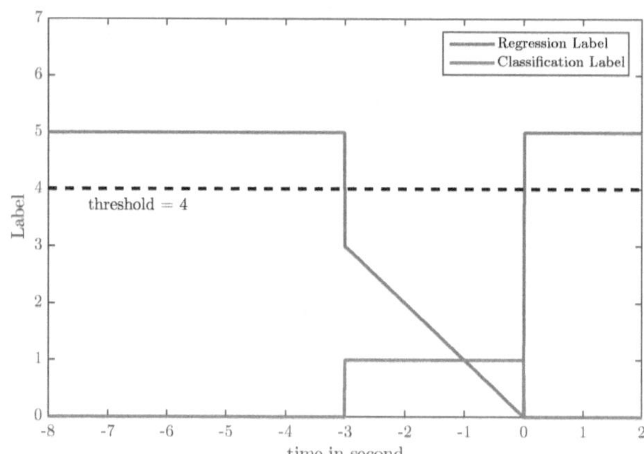

Figure 4.2: Labeling process for classification and regression problems where a lane change occurred at $t = 0$. The negative time line indicates data points that occurred before the actual lane change. Classification labels are discrete, 0 mean no lane change, 1 mean there is a lane change. Regression label are the time left until the actual lane change.

the time left until the driver perform that specific maneuver. For example, the target function for left lane-change is f_{left} such that:

- $f_{left}(X^{(t_0),k}) = i$ seconds, if the driver changes to the left lane in exactly i seconds from the current time step and $i \leq m$

- $f_{left}(X^{(t_0),k}) = m + \Delta$ seconds, if there is no left lane change in the next m seconds. Δ is the offset that allow us to later separate between lane change and lane keeping predictions.

The target function for right lane change f_{right} is formulated analogously to f_{left}.

Setting label of no lane change to $m + \Delta$ seconds is a way of specifying that the model should not concentrate on predicting the correct time-to-lane-change when it is larger that m seconds. In addition, we can set a threshold between m and $m + \Delta$ to convert the regression labels to classification labels. This makes it possible to compare the results of regression approaches with classification approaches.

An example for $m = 3$ seconds, $\Delta = 2$ seconds and conversion threshold of 4 seconds is illustrated in Figure 4.2.

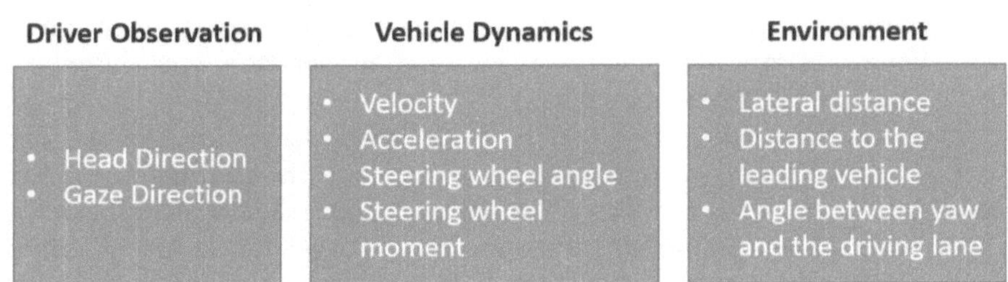

Figure 4.3: Features used for lane change prediction are drawn from three categories: driver, vehicle and environment

4.4 Proposed Approach

To approach the problem of predicting lane change and time to lane change, this section will firstly give an overview of our data set and the input signals. Subsequently, the method for labeling the actual lane changes are described. Furthermore, the design of the proposed network for both lane change and time-to-lane change predictions are introduced and discussed.

4.4.1 Data Set

The data set was collected from 34 subjects with different age and driving experience in a driving simulator. The scenario simulated a two-lane highway with different segments. Each segment was characterized by a specific traffic density (low, medium, high or without traffic) and a specific speed of the leading vehicle on the right lane (90 or 110 km/h). Data of 26 randomly selected drivers (about 75 % of the data set) were used for training our models, and the data from the remaining 8 drivers were used for testing. In total, the training set contains 752 left lane change maneuvers and 753 right lane change maneuvers, and the test set contains 207 left and 206 right lane change maneuvers. With this setting we can evaluate the plausibility of the model when predicting lane change of unseen drivers.

There are in total 9 signals that ware used as inputs for the models. These signals can be divided into three main groups: driver monitoring, vehicle information and environmental information (see Figure 4.3). For the driver monitoring, a multi-camera head-eye-tracking system is deployed, which provides head and gaze directions. The head and gaze movements are converted into numerical time series whose values consist of the relative angles between head (and gaze) direction and the vehicle's longitudinal axis. Vehicle information comprises velocity, acceleration, steering wheel angle and steering wheel moment.

From the environment perception, the following three input signals are used: the relative distance to middle of the lane (lateral distance), the distance to next vehicle in the same lane, and the relative angle between the vehicle's longitudinal axis and the lane. These signals are not only available on the simulator but they could also be extracted from front radar and camera sensors of a real vehicle.

The intention in this work is to not use any information about the traffic situation on other lanes but the current one. The reason being, if the model is given information of the next lanes (i. e. there is someone in the blind spot), it will probably learn that the driver will not make a lane change in such situation. However lane change accidents usually occurs when there is someone in the blind spot and the driver still perform a lane change maneuver. Excluding information of traffic situation on other lanes force the model to learn the driver behaviors to predict lane changes. With that, risky lane change maneuvers can be detected and prevented.

4.4.2 Labeling Lane Changes

Using the lateral distance, a lane change maneuver can be recognized right after it was performed. We can determine the end point of a lane change (the point when a part of ego-vehicle is already on other lane) by finding the jump in lateral distance as in (Lefèvre et al., 2014). There are two types of jump, corresponding to two types of lane change: a jump from positive to negative value indicates that the vehicle changes to the left lane, and a jump from negative to positive indicates a lane change to the right (cf. Figure 4.4). Note that in this approach, the lateral distance measures the distance between the middle of the current lane to the middle of the ego-vehicle. The moment the ego-vehicle touches the lane marking is about 0.5 to 0.8 seconds before the jump.

Figure 4.4 illustrates the labeling process for two lane changes that were performed at about the 398th second and 412th second of a recorded trip. Here the labeling algorithm uses a time window of 3 seconds for labeling the maneuver. The choice of window length for labeling is usually between 2 and 5 seconds.

Based on the lateral distance, one can think about using a simple threshold classify lane changes before it is actually performed. The problem is that at the point when the lateral distance exceeds this threshold, any

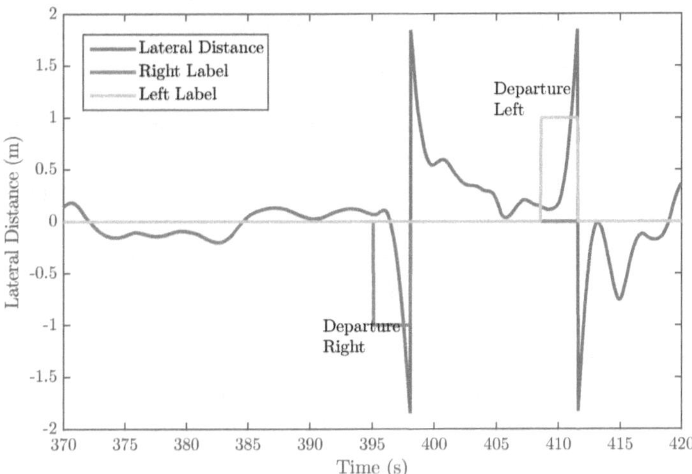

Figure 4.4: Labeling a window of 3 seconds before a lane change. The red signal shows where the data-points are labeled as departure left and the yellow signal shows the right departure labels

warning or active interference from the system could already be too late. In general, the drivers need up to two seconds to acknowledge the warning and perform a correction maneuver. To prevent possible dangerous situations, the system must be able to predict the lane change maneuver a least a few seconds before it occurs.

As already mentioned in Section 4.3, a regression label can be converted back to classification label, but not the other way around. Thus, the regression labels contain more information than the classification labels, and should thus be more difficult to learn than classification one. However, the evaluation on real-world data shows that the learner yields a slightly better result in the regression settings, which has the advantage of providing additional information about the impending maneuver. A deeper look at the evaluation results is provided in the Section 4.5 of this chapter.

Furthermore, time steps that are labeled as lane-change and lie further from the jump are harder to predict but less important than closer ones. In a classification problem, every instance is evaluated equally, which means the cost of misclassifying an instance at one second before a lane change is as same as misclassifying an instance at three seconds before a lane change. To this end, regression formulation is a better metrics for evaluating lane change prediction. Labeling the lane change problem as time-to-lane-change puts more weight on the instances that are chronologically closer to the lane change.

Considering the evaluation of two predictors P_1 and P_2 on a set of two lane change situations. The predictor P_1 is able to predict the first lane change at one second and the second lane change at three seconds before they are performed, and the predictor P_2 predicts both situations two seconds before they are performed. With the classification setting, both P_1 and P_2 are evaluated with an identical score since the total number of corrected predictions is four seconds. However, in the regression setting, P_2 is evaluated with a better score than P_1. From the practical aspect, it could be seen as a trade-off between two good predictions with a very early prediction and one prediction at critical time.

Beyond that, a learner that can predict TTLC gives extra information about the lane change maneuver. On the contrary, when a classifier predicts a lane change, it means that with high probability there will be a lane change being performed soon. For ADAS that changes its warning strategy adaptively based on driver's individual preferences, this information may not be sufficient. For example, elder drivers may need to be warned earlier than young drivers.

4.4.3 Network Architecture

Due to the temporal nature of the driving task and maneuvers, time points in the past may bear significant information for the prediction of upcoming events. Thus, a suitable method for predicting the lane change should process the ability to remember the important information at earlier time steps and discover the dependency between time steps. *Recurrent neural networks* (RNN) are neural network architectures designed for such purposes. RNNs include directed, self-connecting cycles in it design for modeling temporal dependencies. However, training this type of network with long time lag data is very difficult since the gradient could either explode or vanish when it is back-propagated through time.

Long Short-Term Memory (LSTM) is an extended RNN architecture introduced by Hochreiter and Schmidhuber (1997) to solve this problem. LSTM overcomes the above-mentioned problems by introducing memory cells together with gating mechanisms, in which the gates are also learned during training process (cf. Section 2.43).

In this approach, we exploit the power of LSTM layer for capturing the dependency of long time steps. The architecture used in this work

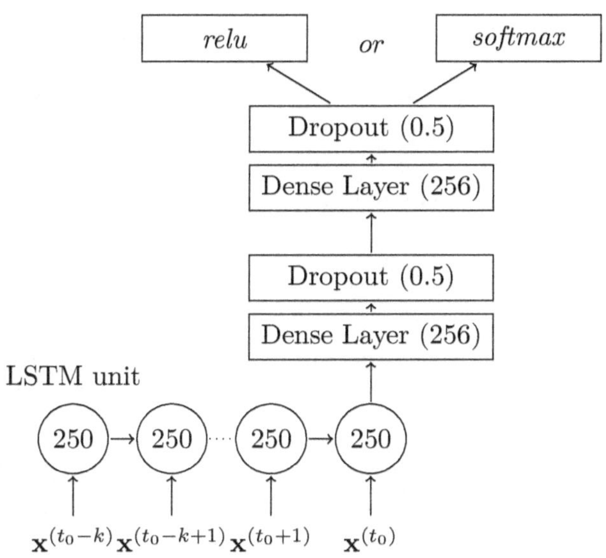

Figure 4.5: Proposed network architecture with the unrolled representation of LSTM layer. The number in each block show how many hidden units are used. The output layer can be either softmax for classification problem and rectified linear unit (relu) for regression problem

consists of an LSTM layer, which is followed by two dense layers and an output layer (Figure 4.5). The specific number of artificial neurons in each layers (hyper-parameters) are found using parameter grid search. For this purpose a small portion of training data is left out as validation set. The best hyper-parameters are selected based on the performance of the model on the validation set.

This architecture is used for both settings, regression and classification, and differs only in the output layer and its activation functions. For regression, two output neurons are used with the activation function rectified linear units (*relu*), one neuron for predicting the TTLC for left departure (f_{left}) and one for predicting TTLC for right departure (f_{right}). For classification, three output neurons are used together with *softmax* as activation function. Each neuron predicts the likelihood that its corresponding action (changing left, right or driving straight) will occur within the next m seconds. Additionally, to accelerate convergence and to prevent overfitting, a batch normalization and a dropout layer with dropping rate of 0.5 are also used after each dense layer (Ioffe and Szegedy, 2015, Srivastava et al., 2014).

For optimizing the regression network, the objective of minimizing the mean squared error (MSE) of TTLC is used. With $y_{i,c}$ and $\tilde{y}_{i,c}$ are the ground truth and prediction for i-th sample of in the dataset. c indicates the direction of the lane change (*left* or *right*). The loss function is defined as:

$$MSE(y, \tilde{y}) = \frac{1}{n} \sum_{i=1}^{n} \sum_{c \in \{left, right\}} (y_{i,c} - \tilde{y}_{i,c})^2 \tag{4.2}$$

The classification network is optimized in the mean of cross entropy loss. Here $y_{i,c}$ indicates the class of i-th sample. c is either *left*, *right* or *straight*:

$$H(y, \tilde{y}) = -\frac{1}{n} \sum_{i=1}^{n} \sum_{c \in \{left, right, straight\}} y_{i,c} \log \tilde{y}_{i,c} \tag{4.3}$$

4.4.4 Personalization

An advantage of neural networks are their ability to reuse a pre-trained network for solving a similar problem on a new data set. In (Zeiler and Fergus, 2014), it is shown that a network that was trained on a bigger data set can be efficiently fine-tuned to cope with a related task, for which there is only a much smaller data set available. It is reported to outperform other state-of-the-art approaches that are only trained on the original data set. Such approaches are called transfer learning, in which, existing models can be re-used for solving a similar or event a new problem. A pre-trained neural network can be re-used in two main ways:

- *Feature construction:* Reuse first layers as a feature extractor and retrain only the last layers for making decision on the new data set. The extracted features can also be used in different learners like random forests or SVM.

- *Initialization:* Using the weights of pre-trained network as initialization for new network and retrain the whole network on the new data set (also known as fine-tuning).

In this work, the second strategy is deployed for fine-tuning the pre-trained networks on data from a specific driver. The LSTM layer is also fine-tuned so that it could adapt to behaviors of new drivers.

The general neural network is herein pre-trained on 26 drivers and this network is then fine-tuned on each of 8 test drivers. Not that in this case, the network does not have to deal with a different problem but rather a different source of data. This approach will be of benefit when the individual behaviors of new drivers are different from what the network has seen in the pre-training process.

Lane-change maneuvers can be detected right after they are performed, and hence they can be used as a ground truth for generating new training samples. From the practical aspect, this means re-training the network can also be done in an online manner. The new training samples can then be collected online for fine-tuning the network. It means that the more often a driver uses this system, the more data will be generated for fine-tuning. The system can therefore adapt itself better on this driver behaviors and provide him with more accurate predictions.

4.5 Evaluation

4.5.1 Evaluation Setup

Since most of the time, drivers move in one lane, the class distribution of lane change problem becomes unbalanced. In fact, a first analysis on the dataset shows that there are only 5.5 % of the driving time are labeled as lane change maneuvers, which means that only 2.7 % of instances are labeled with each of the two lane-change classes. To make the data set become more balanced, down-sampling is applied on the lane-keeping instances so that the number of lane-change instances is about 15 % of the data (7.5 % for each class). As a side effect, this also speeds up the training process which allows us to evaluate more models.

In total, the introduced models are trained on about $360,000$ samples, which are recorded from 26 drivers. The model are validated on $110,000$ samples from 8 drivers. The training and test samples are randomly drawn from two disjoint sets of drivers, so that the generalization capability of models on new drivers can be evaluated.

Baseline models

To compare the proposed LTSM approach with other methods, three other models are also trained on the same dataset. these include a support vector machine (SVM) (cf. Section 2.2.1), and two random forest models (RFs) (cf. Section 2.2.3). One RF uses standard decision trees to solve the classification problem. The second RF uses regression trees to predict time to lane change (regression problem). The configuration of the base line models are as follows:

- *RF Classification* with 200 decision trees.

- *RF Regression* with 200 regression trees.

- *SVM* with a radial basis function (RBF) kernel

The Parameters are determined by grid search. In theory, the higher number of base learner (decision tree and regression trees) the better random forests would perform. However, it also means that there is a higher chance that two base learner will become highly correlated. In fact, in our case, the performance of RFs does not further increase when using more than 200 base learners.

In random forest settings, different feature sampling strategies are also tested. In case of the classification RF, the best number of feature to consider for each split is found to be the around the square root of the total number of features ($\sqrt{n}$). For regression RF, the best feature sampling strategy to sample half of the features ($\frac{n}{2}$)

Evaluation Metrics

Following (Jain et al., 2015), the F1 score used in this book is computed with a small modification to encompass two positive classes (lane change left and right) and one negative class (lane keeping). The true positive predictions (tp) are the correct predictions of either lane change left or right. False positive predictions (fp) are incorrect lane-change predictions and false negative (fn) are incorrect predictions where the lane change actually happened. The precision (PR), recall (RE) and F1 score are formally defined as follows:

$$PR = \frac{tp}{tp + fp} \tag{4.4}$$

$$RE = \frac{tp}{tp + fn} \tag{4.5}$$

$$F1 = \frac{2 \cdot PR \cdot RE}{PR + RE} \tag{4.6}$$

The evaluation is conducted on two networks with the same base architecture (except the output layer) for predicting lane changes and time to lane change. Since the labels for the two problems are different, we cannot directly compare their performance. However, as mentioned in Section 4.4.2, it is possible to convert predictions of TTLC into lane-change classes. For doing so, a threshold τ is used for converting time to lane change to classes:

$$\tau = \frac{m + (m + \Delta)}{2} \tag{4.7}$$

$$= m + \frac{\Delta}{2} \tag{4.8}$$

The threshold τ is actually the average value of the labeling window m and the value used for labeling lane keeping instances $m + \Delta$ (as described in Section 4.3). For example, if the labeling windows of a lane change is $m = 3$ seconds and the Δ is set to 2 seconds then the threshold τ for is calculated as $3 + 2/2 = 4$ (seconds) (cf. Figure 4.2).

To evaluate the TTLC prediction, root mean squared error (RMSE) is used. It is computed as the standard deviation of the prediction errors. Considering evaluating the model on n time steps with the ground truth y_i for $i = 1 \ldots n$. The prediction of each time step is $\tilde{y}_i$. The RMSE metric is computed as follows:

$$rmse = \sqrt{\frac{1}{n} \sum_{i=1}^{n} (y_i - \tilde{y}_i)^2} \tag{4.9}$$

4.5.2 Results of LSTM Networks and Discussion

Lane-Change and Time-to-Lane-Change

In this section, three labeling strategies are evaluated, which have the fixed window length m of 2.5 seconds, 3 seconds and 4 seconds respectively. The results in Table 4.1 show that the regression setting outperforms the classification in all LSTM cases and in two out of three RF experiments. Moreover, the LSTM results are typically better than the corresponding baseline results.

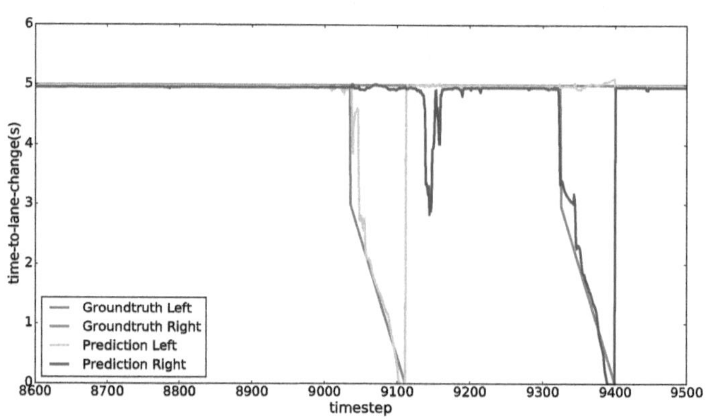

Figure 4.6: Time-to-lane-change prediction of two lane change maneuvers with 3 seconds labeling window

Table 4.1: F1 Scores of baseline models and LSTM networks with different settings for labeling window

Labeling Window	2.5 s	3 s	4 s
SVM	0.896	0.856	0.782
RF Classification	0.893	0.851	0.836
RF Regression	0.909	0.872	0.815
LSTM Classification	0.922	0.892	0.818
LSTM Regression	**0.925**	**0.898**	**0.838**

Using the LSTM regression network, the model is able to predict the time-to-lane-change with an expected error of 0.3 seconds for 3 seconds labeling windows. Table 4.2 shows the Root-mean-squared error of regression model on the testing data set. Figure 4.6 illustrates the predictions of a takeover-maneuver which could be decomposed into two lane-change maneuvers. As we can observe, there are some small peaks in the prediction, which indicates that a filter could be applied on the output prediction to reduce the number of false positive errors (false alarms).

Table 4.2: Root-mean-squared-error of predictions from regression models with different labeling window

Labeling Window	2.5 s	3 s	4 s
RF Regression	0.308	0.342	0.345
LSTM Regression	0.272	0.308	0.329

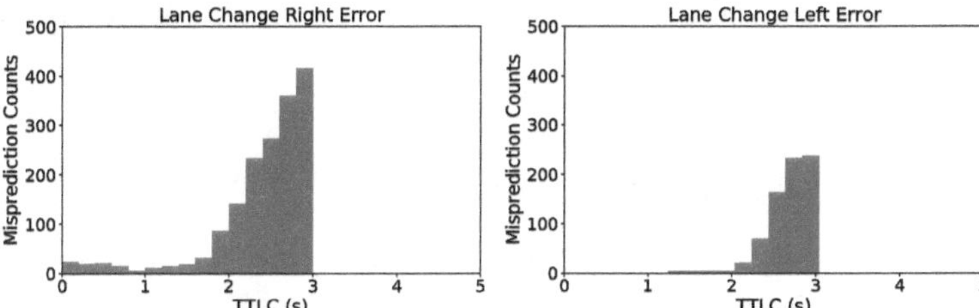

Figure 4.7: Histograms of false negative error for right and left lane changes.

Left Departure vs. Right Departure

From analyzing the results, it could be observed that both networks make fewer mistakes when predicting left departure than right departure maneuver. The histograms of false negative errors in Figure 4.7 show that the model can predict left departure almost perfectly at any time point within 2 seconds before a left lane change. On contrary, right departure still makes a few mistakes within 2 seconds and considerably more mistakes than left departure in the range from 2 to 3 seconds.

In fact, the detailed results in Table 4.3 show that for both settings, regression and classification networks, the false negative error rates of left lane change predictions (7.7 % and 8.4 %) are less than a half of the errors made in right lane change predictions (17.1 % to 19.3 %). The difference in prediction accuracy could possibly be caused by the nature of human behavior when changing to a faster (left) lane and when changing to the slower (right) lane:

- The cause for changing to the left lane is usually a slower leading vehicle and the drivers want to switch to a faster lane and overtake the leading vehicle. Since the distance to leading vehicle is also included in the inputs, the network learns this behavior and is able to predict a left-change maneuver earlier and more precisely.

- To perform a left change maneuver, drivers usually check the side mirror more carefully (and maybe several times) than when changing to the right (slower) lane. This behavior can be captured in the signals of head and gaze direction. A deeper analysis on the these two signals can confirm that observation. The average time that a driver spends looking at the left side within five seconds before changing to left lane is about 1.3 seconds whereas the time drivers spends looking on the right before changing to the right lane is only about 1.0 second.

Evaluation of False Negative Error

As mentioned above, misclassifying an instance that is closer to the lane change could be more costly for an ADAS. With the here proposed time-to-lane-change approach the network makes fewer errors within 2 seconds before a lane change. Table 4.4 shows the number of misclassified lane-change instances before and after 2 seconds for both regression and classification network. 7.47 % of the total false negative errors made by regression network are fall into 2 seconds before a lane change whereas with classification network, the ratio is 11.12 %.

Result of Personalization

To evaluate the personalization of the model, the data from 8 test drivers that have not been seen during training are used. For each of them, the data is split into two parts, the first 20 minutes are used for fine-tuning the general model and the last 10 minutes are used for testing the performance of the personalization. Note that the number of training data used for fine-tuning the personalized model is relative small compared to data used for training the general model.

Table 4.3: Classification Errors of Right and Left Departures

	Samples	Correct	False	
Class. Right	8676	7002	1674	(19.29 %)
Class. Left	8840	8095	745	(8.43 %)
Reg. Right	8676	7191	1485	(17.12 %)
Reg. Left	8840	8155	685	(7.75 %)

Table 4.4: Number of prediction errors before and after 2 second mark

	Misclassification	**> 2 s**	**< 2 s**
Classification	2419	2150 (88.87 %)	269 (11.12 %)
Regression	2129	1972 (92.63 %)	157 (07.37 %)

Table 4.5: F1 score Comparison of individual and general models on different settings

	Regression			**Classification**		
Labeling Window	2.5	3	4	2.5	3	4
General Model	0.925	0.903	0.803	0.921	0.894	0.806
Personalized Model	0.929	0.905	0.816	0.923	0.901	0.815
Δ F1	0.003	0.002	0.013	0.002	0.007	0.009

To prevent overfitting on the individual model, the learning rate is reduced to 1/100 of the learning rate used for training the general model. Table 4.5 shows the average F1 scores evaluated on 8 test drivers using the general model and the personalized models. The results show improvements on both regression and classification settings, where the improvements are higher for settings with larger labeling window (larger m). That indicates the benefit of the personalized model over the general model for predicting a lane change at earlier timesteps. This results could also be interpreted as evidence that the behavior of test drivers up to 2.5 seconds before lane change are similar to behaviors learned from training set, whereas from 3 to 4 seconds before a lane change, the behaviors of test drivers are more individualized.

4.6 Conclusion

In this chapter, we formulated the problem of lane change prediction as time-to-lane-change prediction, which means changing the perspective from a classification to a regression approach. We compared both approaches using the same long short-term memory (LSTM) architecture with three baselines and showed that LSTM networks deliver better prediction performance than the baseline models in both cases. Furthermore, despite dealing with a harder problem, the LSTM regression network is able to provide comparable results with classification setting of lane-change prediction. In addition, a regression approach can provide extra information about the exact time of the impending the lane change for high-level applications of advanced driver assistance systems. Finally we pointed out that fine-tuning a general network model on only 20 minutes of driving data of a specific driver leads to a better prediction performance for this driver.

5 Personalization by Exploiting the Dependency between Maneuvers

5.1 Introduction

It is often the case that users have to adapt themselves to a new system or function to be able to use it. When the users' expectation and preference of a system are not met, their trust in the system will decrease and eventually they may ignore or turn it off. For a standard not-adapting system, there could be a gap between the user's expectation and the outcome of the system. Personalization systems aim at closing this gap by adjusting themselves to the current user.

The problem of personalization is already addressed in different field studies and applications. Recommender systems are an example, in which personalization play a central role to improve prediction performance. In recommender systems, the data are mostly defined as a sequence of actions. For instance, the action could be "visiting a website","buying an item" or "clicking on an ad" (Aggarwal, 2016). The problem of personalization in this case is usually formulated as a prediction task in which the model is given an action A_{i-1} or a sequence of actions $A_{0..i-1}$. Based on that, the system must then predict the next action A_i that the user is likely to perform. In other words, it learns a function $\pi(\cdot)$ to forecast the future actions:

$$\pi : A_0...A_{i-1} \rightarrow A_i \tag{5.1}$$

In the automotive field, the sequence of maneuvers may by itself reveal very little about the driver. For example, if a driver wants to get to his travel destination, certain maneuvers have to be performed in any case (i.e. entering a roundabout, turning at an intersection and so on...). However, different drivers vary in the way they perform these maneuvers. As a consequence, the problem of personalization must be reformulated to: based on observing how a driver performs the driving maneuvers to infer

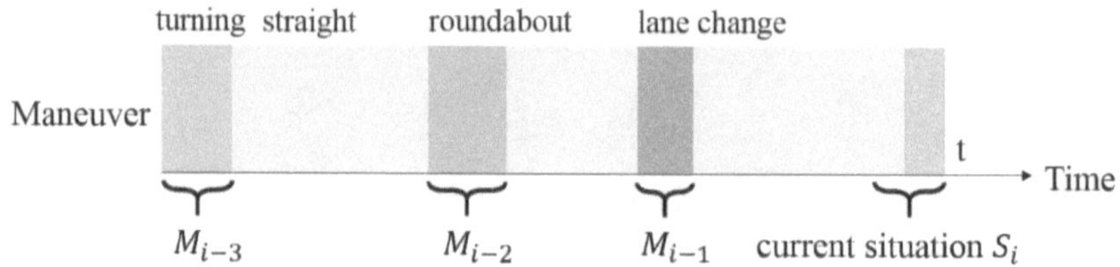

Figure 5.1: A route is separated into driving maneuvers

the driver's preference or to predict how she would perform the next maneuver or some specific maneuvers of interest. In other words, we have to introduce a function $f(\cdot)$ that extract driving styles from maneuver execution. The personalization problem now become to learn a function $\pi(\cdot)$ which maps the driving style $f(.)$ of previous maneuvers to the driving style of the upcoming maneuver:

$$\pi : f(A_0)...f(A_{i-1}) \rightarrow f(A_i) \tag{5.2}$$

Understanding the driver is a challenging task, there are many factors that can influence a driver on her decision making process and also on how an action is going to be performed. These influencing factors include the driver's habit, her driving skills, her experience, her current physical and mental state and so on. Observing all these factors are complex, and modeling how they will affect the driver and the way she drives is not a simple task.

In this work, driving trip is considered as a sequence of maneuvers, interleaved with periods of default activities such as driving straight (see Figure 5.1). With that we can treat the maneuver executions as samples for training machine learning models. To tackle the problem of personalization, this chapter presents a new approach by learning the dependency between maneuver executions. We formalize the problem and show that it is reasonable to exploit the dependency between maneuver executions to adapt the model to individual drivers.

The remainder of this chapter is structured as follows. In the next section, Section 5.2, other related approaches in personalization are reviewed. Section 5.3 gives an overview of the data set and the use case that we are going to solve.

We firstly tackle the problem by revealing the dependencies between maneuver executions using unsupervised learning in Section 5.4. Subsequently, based on the results of the unsupervised learning approach, we can further improve the approach by using supervised learning to learn the dependencies between maneuver execution. The supervised approach is presented Section 5.5. Section 5.6 dives deep into how the proposed approach can be implemented using neural networks. The evaluation of the prediction performance is then presented in Section 5.7. Finally, Section 5.8 concludes this chapter and discusses future steps.

5.2 Related Work

The idea of personalization is to enable the system to adapt itself to the driver and thus is a key to further improve the safety and driving comfort. The advantages of personalization have already been observed in different applications in automotive field, such as adaptive cruise control (longitudinal driving assistance) or lane changing prediction (lateral driving assistance). In this section, a review of methods for personalization in literature is given. Furthermore, to provide an overview of what data can be used, the categories of features used for personalization applications in automotive fields are also reviewed.

5.2.1 Methods for Personalizing ADASs

Personalization is often combined with the idea of driving style recognition, since both problems address the same issue of understanding the driver behavior. One of most common approaches is to divide drivers in different groups (e.g., aggressive, moderate and calm). The number of groups varies from two to six, depending on the particular application of interest. The group information could either be used directly to assign individual recommendation to driver or it could also be used as input to the personalization of some specific functionality of ADAS.

In this work, we use clustering method to analyze three different maneuvers. Based on the correlation between the cluster result and the gap acceptances, our system can quickly provide driver with personalized recommendations for the left turn situations even when the driver hasn't yet performed any left turn maneuver.

Rosenfeld et al. (2015) proposed to use demographic input to personalize the prediction of driver's preference and showed a improvement over non-personalized models. However, drivers would need to enter their demographic information to receive personalized information. In the practical point of view, it would be a disadvantage. Further demographic is mostly static and cannot be used to predict intra-individual preferences of drivers.

To personalize a machine learning model in general, one can firstly learn a general model from the collected data. We call this a general model since it works well for an average user/driver. The personalization process then adapts the model to a new driver by fine-tuning it on data collected from that driver (Dang et al., 2017). This method requires the underlying model to be able to learn in an online or batch setting. The problem of overfitting should also be considered while using this approach since only a small portion of data from the new driver is available.

In many applications in the automotive field, the personalization problem can be reduced to identifying the individual gap acceptance of the driver. Butakov and Ioannou (2015) propose an approach to personalize the lane change prediction by modeling the longitudinal adjustment behavior and the gap acceptance of driver. The gap at a lane change maneuver is defined as the space between the leading vehicle and the following vehicle on the target lane. For adaptive cruise control, an adaptive system will automatically adjust the distance to the leading vehicle to match the driver's preference. Another application of gap acceptance is to predict the decision of the driver at an intersection where driver has to wait for a appropriate gap to perform a turning maneuver (Cooper and Zheng, 2002).

5.2.2 Personalization Features

Driving style is an abstract concept that is not directly observable from the data. What we can observe is the reflection of driving style on some measurable signals like speed, acceleration, etc. A survey conducted by (Martinez et al., 2018) shows that the choice of input signals to use for detecting driving style variates from application to application. In general the input signals could be divided into three main groups:

- Vehicle dynamics: i.e. speed, acceleration, jerk, etc.

- High level behavioral features like turning, car-following, distance-keeping, etc.

- Power or fuel consumption.

- Personality traits or demographic data.

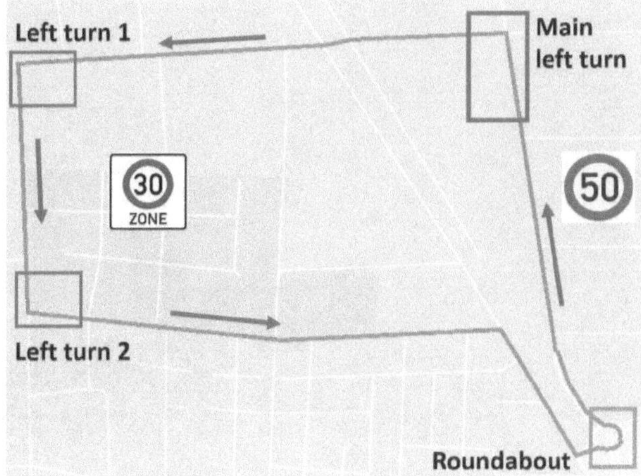

Figure 5.2: Test route used for recording data. The roundabout and left turn situations are highlighted with orange and red boxes

In this work we focus on using the vehicle dynamics information packed in the form of maneuver executions. The reason being this kind of features is always available in all maneuver executions and they directly reflex the driver's preferences and driving styles. Information like power or fuel is more specific for optimizing the energy consumption. Whereas the information such as personality traits is not available without firstly conducting psychological tests on drivers.

5.3 Data Set and Application

5.3.1 Data set

The data used in this work were collected from 32 drivers which cover a wide range of ages and driving experiences. About 31.3 % of the drivers are 26 years old or younger, 43.8 % are between 27 and 60 years old, 25 % are over 60 years old. The annual mileage of each driver varies from few hundreds to 50,000 km.

From each driver we recorded 30 rounds of driving on a pre-defined route. The data set was collected using the PRORETA vehicle in real-world traffic. The vehicle is equipped with different sensors for capturing driving dynamics and ego-position and other traffic participants (type,

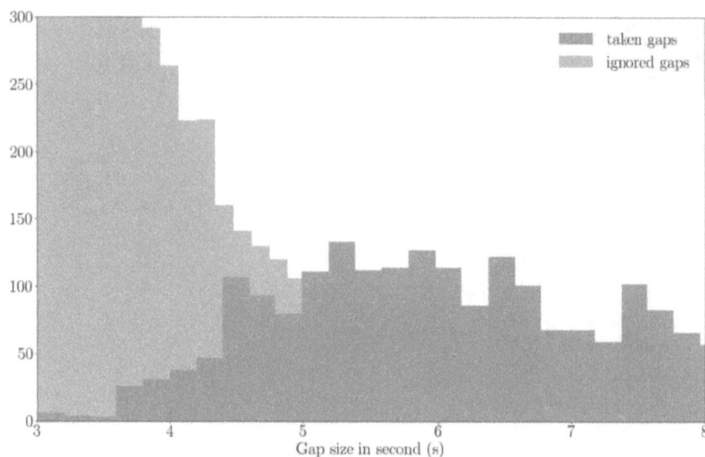

Figure 5.3: Histograms of taken and ignored gaps recorded from all 32 drivers at the left turn scenario

position, speed ...). The chosen route is a common urban route which requires the driver to perform different maneuvers like driving through a roundabout, make a left turn, or pass an intersection with left yields to right, etc. To complete the trip, each driver has to perform different maneuvers per round. To evaluate the idea of exploiting the dependency between maneuvers, we will focus on the two most complex maneuvers, roundabout and left-turn maneuvers. It is also reasonable to choose complex maneuvers since they require more actions from the driver and are therefore more personalized to each driver. An illustration of the route is shown in Figure 5.2, where the roundabout and left-turn maneuvers are highlighted with an orange and a red box, respectively. The evaluation of prediction performance is only conducted at the main left turn since it has high traffic volume. The other maneuvers (roundabout, left turn 1 and 2) are used as previous maneuvers.

For capturing the driving style, four signals are used as features in each maneuver execution, which include speed, longitude acceleration, latitude acceleration and steering wheel speed. As features for describing the current left turn situation the gap size in second, velocity and distance to intersection of ego-vehicle

5.3.2 Gap Acceptance at Left-Turn Maneuver

In an unsignalized intersection scenario, in which the incoming traffic has the right of way, the driver has to wait for a suitable gap to turn left. We can observe that the preference of the driver in choosing gaps differs from driver to driver. Gaps that are in range from approximately 3 to 7 seconds could either be taken or ignored, depending on the driver's preference, as shown in Figure 5.3. In such a scenario, the problem of predicting the gap acceptance of driver should not only consider the current traffic situation but also the driver's individual preferences.

While the ego vehicle is approaching the intersection, we can detect the oncoming traffic using the equipped front-radar, as well as the relative position, speed and size of the cars. Based this information, each potential gap for turning left is computed. An illustration of the application at an intersection is shown in Figure 5.4, where the ego vehicle is depicted as a red box and the incoming vehicles are shown in gray. The size of the first gap is the distance between the closest incoming vehicle and the middle of the intersection (reference line). The last gap is computed based on the furthest detected incoming vehicle and the maximum radar range. All gaps between these two gaps are bounded by a leading vehicle and a following vehicle. For each gap i we compute the time (τ_i) that is available for the driver to turn left:

$$\tau_i = \frac{s_i}{v_i} \tag{5.3}$$

where the s_i is the longitudinal size of gap i and v_i is the current speed of the following vehicle that forms gap i. If there is no following vehicle detected, the following speed of that gap is set to the maximal speed allowed on that street plus 10 %.

5.4 Maneuver Clustering Approach

5.4.1 Motivation

The disadvantage of fine-tuning a model is that it needs a certain amount of data to be able to efficiently update the base model. That means if we want to personalize a support system for left turn scenarios, the model would have to firstly record multiple left turn maneuvers from the new driver. The driver would therefore have to wait for days or event weeks before she can experience the personalized supports. Furthermore, if the driver changes her driving style, e.g. when she answers a phone call and shifts to

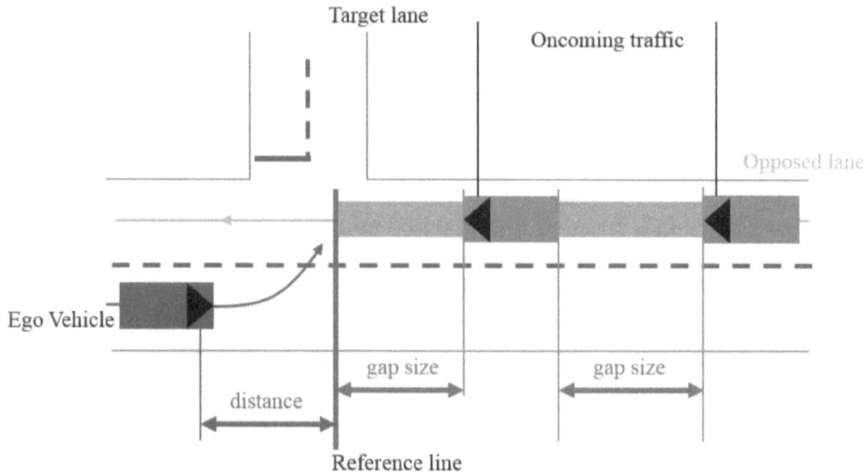

Figure 5.4: A left turn situation with two possible gaps

a more relaxed driving style, the system will need to observe even more left turn maneuvers before being able to adjust the model accordingly. The idea of exploiting the dependencies between maneuver executions can be used to quickly discover such changes in driving style and thus solve this problem.

In the context of personalization, assigning the driver into different groups (driving styles) has been used in various automotive researches and applications (cf. Section 3.2). In this section, we combine the idea of exploiting the dependencies between maneuver executions and driving style recognition. This approach make use of clustering methods to discover the driving style clusters from the data. We then analyze the dependency between the cluster assignment and the gap acceptance of each cluster. In particular, for each cluster, the acceptance curve is computed based on the statistics of actually taken and ignored gaps modeling the probability that a driver will take a gap of certain size. The driver's individual acceptance curve can then be updated each time the driver executes a maneuver.

5.4.2 Maneuver Clustering

In total, three different maneuvers are used for the clustering approach: driving through a roundabout, approaching an intersection, and turning

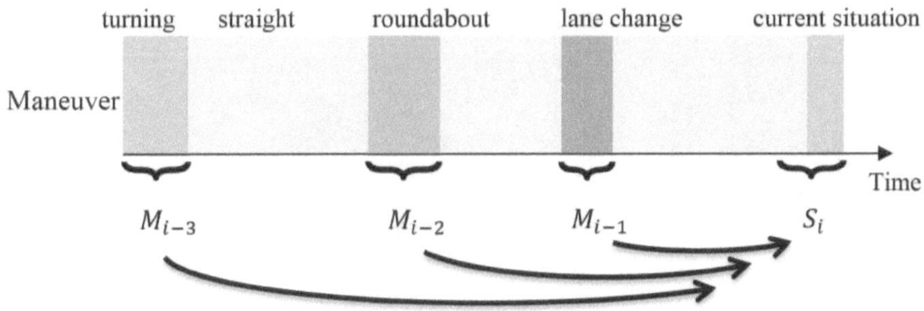

Figure 5.5: Incorporating past maneuver executions to personalize the prediction of current situation. M_{i-1} is the time series representing last maneuver, and S_i represent the current situation (traffic, gaps, etc.)

left. Each maneuver execution is recorded as a time series of vehicle dynamics. In particular, each maneuver contains the sensor values over time of velocity, longitudinal and latitudinal acceleration, yaw rate, jerk rate and steering wheel speed. The statistical values from each signal are computed and used as features for clustering the maneuver executions.

At the training stage, the K-means algorithm is used to separate the maneuver executions of each type into three groups (cf. Section 2.1.1). Based on the statistics of actually taken and ignored gaps, the gap acceptance is computed for each group and each maneuver type. To compute the probability of a gap being taken two assumptions are made:

(a) If a driver takes a gap of size $\tau - \epsilon$ seconds, $\epsilon \geq 0$ then she will also take a larger gap (e. g. with size of τ seconds).

(b) If a driver ignores a gap with size $\tau + \epsilon$ then she will also ignore smaller gaps with size τ.

These two assumptions allow the probability of a gap with size τ to be computed even if there is no data of the exact gap size of τ seconds. The probability of a gap with size τ being taken is:

$$P(\tau) = \frac{\sum_{i=1}^{n}[g_i \leq \tau]}{\sum_{i=1}^{n}[g_i \leq \tau] + \sum_{j=1}^{m}[h_j \geq \tau]} \tag{5.4}$$

with n and m being the total number of taken and ignored gaps respectively and $[\mathbf{x}]$ being the counting function that returns 1 if $\mathbf{x}$ is true and 0 otherwise. g_i and h_j are the size (in seconds) of taken and ignored gaps respectively.

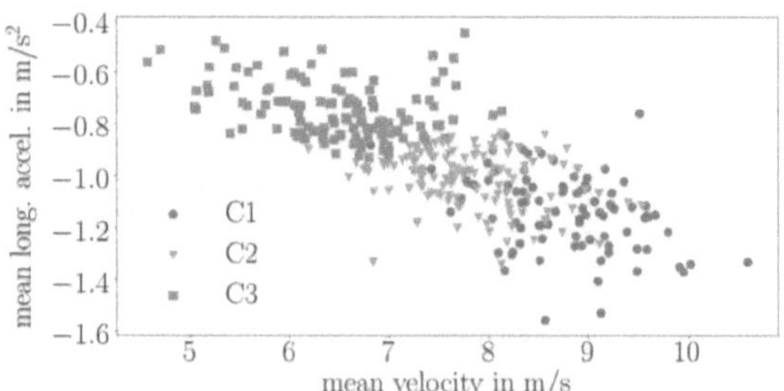

(**a**) Clustering the of maneuver execution. Each dot represent a maneuver execution. The color shows the cluster assignment

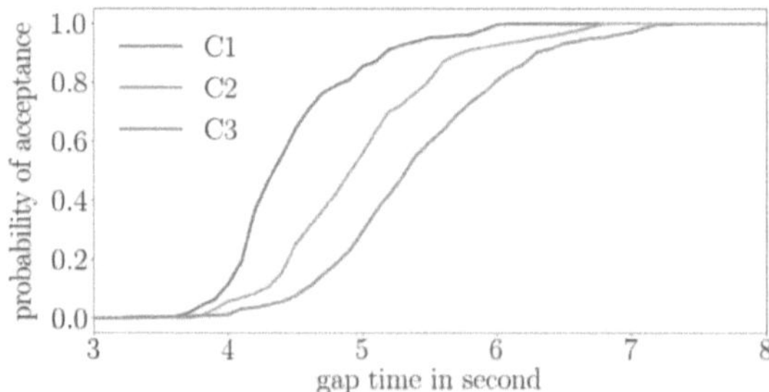

(**b**) The corresponding acceptance curve for each cluster computed based on the taken and ignored gaps.

Figure 5.6: Visualization of clustering results and the acceptance curves

Figure 5.6a shows the clustering of the maneuver executions of approaching an intersection. The $x-$axis shows the mean of velocity and the $y-$axis shows the mean of longitudinal acceleration. Each dot is a recorded maneuver and each color shows the cluster assignment. Note that the figure is created in two-dimensional space but the actual clustering algorithm is performed in a higher dimensional space which includes further vehicle dynamic features as mentioned above. The cluster $C1$ contains maneuver executions with higher velocity and high longitudinal deceleration when approaching an intersection whereas $C3$ contains more defensive behaviors with lower velocity and lower longitudinal deceleration. $C2$ shows a balanced style between $C1$ and $C3$.

The Figure 5.6b shows the corresponding acceptance curves for each cluster computed from the taken and ignored gaps of a left turn within the same short recording as the respective maneuver. The acceptance curve specifies the probability of accepting a gap given the gap's size, here separated for each driving style cluster. Therein, the correlation between the clustering assignment and the acceptance curve can be observed. The acceptance curve of $C1$ is on the left side in comparison to $C2$ and especially $C3$. This means $C1$ usually accepts smaller gaps than $C2$ and $C3$. On the other hand, $C3$ is more defensive and mostly takes larger gaps than $C1$.

Computing the Accumulated Acceptance

When applying the system on the vehicle, whenever the driver executes one of the three maneuvers, her cluster assignment is computed and the corresponding acceptance curve is selected and used to gradually update her individual acceptance curve. For this, the exponentially weighted moving average (EWMA) is used to accumulate the acceptance curves computed from executed maneuvers.

$$\overline{P}_i = (1 - \lambda)\overline{P}_{i-1} + \lambda P_i \tag{5.5}$$

$$= \overline{P}_{i-1} + \lambda(P_i - \overline{P}_{i-1}), \tag{5.6}$$

where P_i is the current acceptance curve computed at maneuver i, $\overline{P}_{i-1}$ and $\overline{P}_{i-1}$ are the accumulated acceptance value before and after seeing maneuver i respectively. The parameter λ specify how much the last maneuver execution contributes to the final acceptance curve. This update process allows the system to gradually forget the older maneuver executions and put more weight on the newer ones where the weighting parameter controls the momentariness, or respectively the persistence, of the driving style model.

5.4.3 Discussion

Figure 5.7 summarizes the whole workflow of the unsupervised approach. Note that, the gap acceptances are only computed after K-means has discovered the three driving style groups and K-means only use vehicle dynamics as input. The discovered clusters are therefore not bounded to the purpose of predicting gaps but they can be used for other purposes.

Furthermore, by observing the distribution of each clustering along specific feature dimensions, we can even match the cluster to driving style categories that usually used in the literature. For example, we can match C1 with the sporty/aggressive driving styles, C2 with the normal/calm

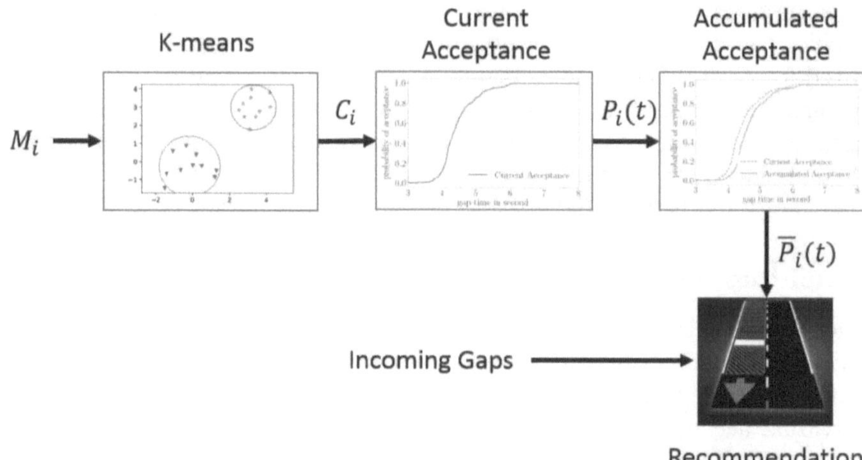

Figure 5.7: Workflow of clustering approach for updating a new maneuver

driving styles and C3 with the defensive/anxious driving styles. Such information is not only helpful for personalizing the assistance systems but it could also be used as feedback for driver. Further, the results of clustering algorithm in this case is transparent and interpretable to human. Which means it can be used as an explanation for the recommendations produced by the system.

5.5 Supervised Approach

5.5.1 Problem Formulation

Maneuver Execution

The sensors integrated in the current cars enable us to recognize different driving maneuvers and to capture the vehicle dynamics during the maneuvers. In this section, we will first formalize a maneuver execution and integrate this on driver assistance system to improve and personalize the prediction.

Given a situation S_i and a driver D_j, the driving maneuver could be formulated as a function of S_i and D_j:

$$M_i = f(S_i, D_j + \epsilon_i) \tag{5.7}$$

where S_i represents all environmental factors, such as traffic situation or weather, whereas D_j constitutes the driver's dispositional factors. Both S_i and D_j have an impact on how a maneuver will be performed. However, the same driver could behave differently even if the same situation repeats at some other time. Thus, a variable ϵ_i is introduced to capture the fluctuations in the driver's behavior. It can be seen as a noise term on driver's impact while performing maneuver M_i. We also have to keep in mind that D_j are a set of different factors like driving skills, driver's experience, driver's habit and so on. Detecting these factors and identifying how they influence a maneuver execution (M_i) are complex tasks and hard to model. Because of that, these factors are usually not considered in current driver assistance systems.

Incorporating Previous Maneuvers

In general, we want to predict whether or not a driver will perform a specific action or how the driver will preform this action. As inputs, the current state of the ego vehicle and the information about the current traffic situation are commonly used. The prediction task in this case can be formalized as: given the current situation S_i and a target value (label) y_i, learn a function g, that predicts y_i based on S_i.

To personalize the predictions of y_i, we additionally have to consider the influences of the driver $D_j + \epsilon_i$ in this situation, since y_i depends on both factors:

$$y_i \leftarrow S_i, D_j + \epsilon_i \tag{5.8}$$

Now we have the same problem as in modeling maneuver execution, namely that the driver's dispositional factors, which influence y_i, are not directly observable. This make it impossible to learn the individual impacts of driver D_j on y_i.

In general, the approach to adapt a model to the new data source (in our case the adaptation of the prediction of y_i to a given driver) is to collect data from the new source (D_j), and learn a new model from it. Alternatively, the existing model could also be reused as a starting point for fine-tuning on the new data (Dang et al., 2017) or for learning a correction model that patches the existing model (Kauschke and Fürnkranz, 2018). However, this process requires a sufficient amount of data so that the new model can adapt itself to the new data source.

As mentioned above, in real-world driving data, the drivers' factors are encoded in almost each maneuver execution. That mean each maneuver in the past could contain valuable information about the driver. The results of clustering analysis shows that there is a dependency between maneuver executed by a same driver. The idea is therefore, instead of learning y_i as a function of S_i and D_j, we can learn y_i as a function of S_i and M_{i-1}. Here M_{i-1} is a maneuver that was recently performed by the same driver D_j. This temporal restriction of M_{i-1} allows us to approximate the current impact of the driver $D_j + \epsilon_i$ with $D_j + \epsilon_{i-1}$, which was captured in the previous maneuver M_{i-1}.

$$\begin{aligned} y_i &\leftarrow S_i, M_{i-1} \\ &\leftarrow S_i, f(D_j + \epsilon_{i-1}, S_{i-1}) \\ &\approx S_i, f(D_j + \epsilon_i, S_{i-1}) \end{aligned} \tag{5.9}$$

The prediction of y_i can then be written as a function G of S_i, $D_j + \epsilon_i$ and S_{i-1}:

$$\begin{aligned} y_i &= g(S_i, f(D_j + \epsilon_i, S_{i-1})) \\ &= G(S_i, D_j + \epsilon_i, S_{i-1}) \end{aligned} \tag{5.10}$$

By taking M_{i-1} into account, we are also using the previous traffic situation S_{i-1} for predicting the current situation. With the assumption that y_i does not depend on other situations than the current S_i, there should be no information of y_i contained in the previous situation S_{i-1}.

$$P(y_i|S_i, D_j) = P(y_i|S_i, D_j, S_{i-1}) \tag{5.11}$$

This means that with enough training data, the learner G could learn the independence between y_i and S_{i-1} and extract the driver impact D_j from M_{i-1}. In other words, G will be forced to learn to extract useful information about the driver's style or behavior from past maneuver executions, and then learn how they will affect the behavior of the same driver in the current situation.

5.5.2 Extracting Driver's Information

A maneuver execution (M_i) is characterized by a multivariate time series of sensor values (see Figure 5.1):

$$M_i = (x_i^{(1)}, x_i^{(2)}, ..., x_i^{(n)}), \quad n > 0 \tag{5.12}$$

where n is the length of the time series and $x_i^{(t)}$ is a column vector of sensor values at time t. It has to be noted that the length of a maneuver varies, depending on the driver and the traffic condition. To learn from this kind of data we either have to convert them into a fixed-length feature vector or make use of a model that can handle time series with variable lengths.

In this section, we are going to examine two approaches for extracting a driver's information from the last executed maneuvers. The first approach extracts features from the time series by computing the statistical information like minimum, maximum and standard deviation from each series of sensor values. The second approach makes use of a recurrent network layer and takes the whole maneuver execution as an input sequence. While the idea of the first approach is widely used in the literature for extracting driver's information, the second one does not make any assumption about the statistical values but takes the raw data and learns to extract useful information for the prediction.

5.5.3 Overfitting

We also need to be aware that using recently performed maneuvers M_{i-1} as features to predict y_i could potentially lead to overfitting because part of the input (i.e., S_{i-1} which is contained in M_{i-1}) is irrelevant to the prediction of y_i. The more additional maneuver executions we take into account, the more likely our model will overfit the training set and thus decrease the prediction accuracy on unseen situations on the test set.

Straight-forward approaches to deal with this problem are to increase the training data and to use regularization techniques. In this work, we create augmented data by adding small artificial noise to the original data. Also regularization techniques like dropout (Srivastava et al., 2014) and batch normalization (Ioffe and Szegedy, 2015) are also deployed to reduce the effect of overfitting. Furthermore, we also apply early stopping to monitor the training process. The training process will be terminated if the validation score does not increase in the next epochs (Prechelt, 1998).

5.5.4 Intra-Personalization

In the problem formulation section (cf. Section 5.5.1), the term ϵ is introduced to specify the intra-personal differences of the driver. To be able to model the intra-personalization we need to assume that ϵ is stable in a certain time window and does not randomly switch from one state to another. In other words we assume that the driver's dispositional factors like driving skills, driving style and preferences stay the same between two consecutive maneuvers or maneuvers that are performed in a short time window. This assumption may not hold for other factors like emotion, but to assess that, we may need more sophisticated sensors for monitoring the driver.

With the assumption above, the problem of intra-personalization could also be tackled by exploiting the dependency between maneuver executions of the same driver. The model can adapt to the changes in recent maneuver execution and use that for anticipate the future maneuvers. However validating intra-personalization would require a bigger data set that is recorded from the same drivers over long period of time.

5.6 Modeling Maneuver Dependency with Neural Networks

We have now formulated the prediction of y_i as a function of the current situation S_i and the previous maneuver execution M_{i-1}. By incorporating the previous maneuver M_{i-1} as input, the individual influence of the current driver will also be considered for predicting y_i. We can extend this concept by incorporating k maneuvers that were performed by this driver. Figure 5.8 shows an example of a network that uses $k = 3$ other maneuver executions in addition to S_i. To validate this concept, this section evaluate it with $k = 1$ which is theoretically easier to train and requires less data. We use a neural network to model the function G, whose objective is to predict the taken gaps.

5.6.1 Network Architectures

We design the network using two input layers separately, one layer takes the current situation (S_i) as input and the other layer is used for capturing useful information from the previous maneuver (M_{i-1}). Each of these two input layers are then followed by hidden layers and form two separate paths. For the ultimate purpose of predicting y_0, these two paths are then combined in the deeper layers of the network and then followed by further hidden layers and lastly the output layer. The whole network is trained by back-propagating the classification error.

Using this network structure allows us to customize the two input paths individually and also make it possible to apply different regularization strategies on each path. This is useful to deal with overfitting since it could

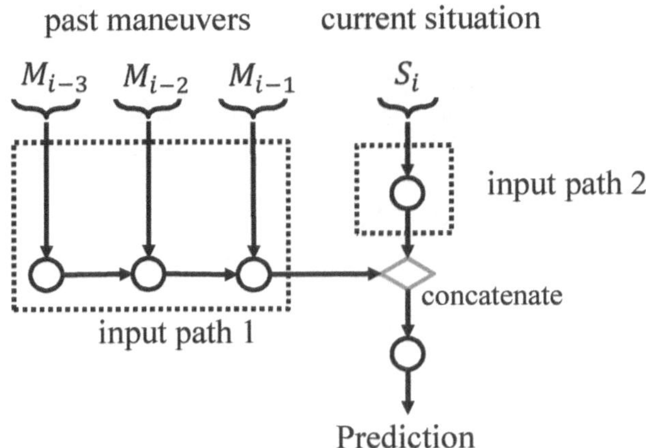

Figure 5.8: Proposed architecture to capture the driver information from previous maneuver executions and use that for predicting the current situation

be a problem when using previous maneuvers as features. Furthermore, the input path for capturing M_{i-1} could also be removed from the architecture, which results the common approach that learn to predict y_i directly from S_i.

5.6.2 Extracting Driver Information with Neural Network

As mentioned in Section 5.5.2, extracting driver information from M_{i-1} could be done in two different ways. We modeled the first approach by configuring the first input path using fully connected layers that take statistical values computed from the last maneuver execution as input. The function G could then be written as $G(S_i, h_s(M_{i-1}))$ or $G_s(S_i, M_{i-1})$ for short. Here h_s compute the statistic information of the given maneuver. The statistic features used for this case are the same feature used in unsupervised approach (cf. Section 5.4).

The second approach makes use of a recurrent layer to capture the whole maneuver execution. Long short-term memory networks (LSTM) (Hochreiter and Schmidhuber, 1997) have proven to be successful for sequence learning problems (Dang et al., 2017, Sutskever et al., 2014), so we use them in this work capturing and propagating information of the

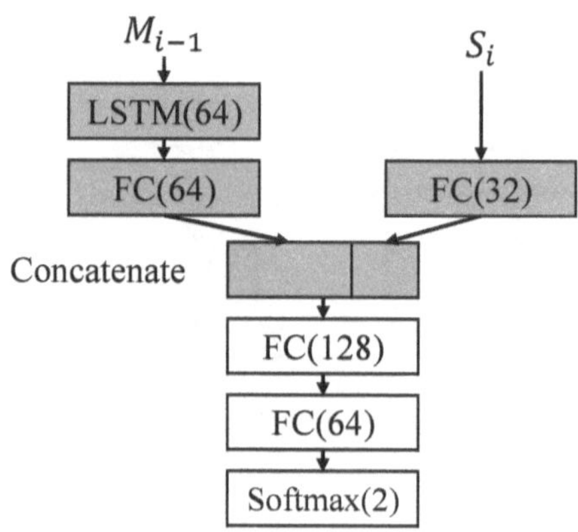

Figure 5.9: Proposed network using LSTM layer for capturing maneuver data. Each fully connected layer (FC) is followed by a dropout layer. The number in each layer depicts how many neurons are used.

individual past maneuvers M_{i-1}. The function G in this case is formalized as $G(S_i, h_{lstm}(M_{i-1}))$ or $G_{lstm}(S_i, M_{i-1})$ for short. Here h_{lstm} depicts the LSTM layer that return hidden representation of the given maneuver M_{i-1}. The detail architecture of this approach is shown in Figure 5.9.

5.7 Evaluation

5.7.1 Evaluation Setup

Cross-Validation

The evaluation the proposed method is conducted using 10-fold cross-validation to construct training and testing sets. The data are split into 10 equal and disjoint folds. Two splitting strategies are evaluated (see Figure 5.10):

- *Splitting according to driver*: In this setting, we divided training set and testing set according to driver so that the training and testing data originate from two disjoint sets of drivers. This assures that the model will be trained and tested on data collected from different drivers.

- *Splitting according to round*: In the second setting, the training and testing sets are divided according to the round id. This setting still assure that training and testing sets are disjoint in terms of rounds (no round will appears in both sets), but it allows the training and testing sets to contain data from the same drivers.

With the first setting, we can approximate the expected score of the system on new drivers that the tested model has not seen before, whereas the second setting focuses on estimating the performance for known drivers. However, the results of conducted experiments did not show substantial differences between the performance estimates obtained with both settings, so in the following analysis, we will only discuss the results for the first setting.

Evaluated Models

In total we evaluate four models. As the baseline model, we can estimate the best possible threshold θ that separates the taken gaps from the ignored gaps in the training set. The prediction the threshold model for a gap with size $= \tau$ seconds is simply *true* (gap will be taken) if $\tau > \theta$ and *false* otherwise. The baseline is compared to the three models described in the previous sections:

- $G(S_i)$ considers only the current situation as input (the path that connects to M_{i-1} is removed).

- $G_s(S_i, M_{i-1})$ uses the statistical values from the roundabout maneuver

- $G_{lstm}(S_i, M_{i-1})$ captures the whole maneuver execution sequence M_{i-1} in an LSTM layer.

Evaluation Scores

In a dense traffic situation, the driver usually has to ignore several small gaps before she finds a suitable one. This fact is also reflected in the distribution of taken and ignored gaps. About 31 % of the gaps are taken. Since the classes are slightly imbalanced, we make use of F1 score in addition to the accuracy score to evaluate the performance of the models.

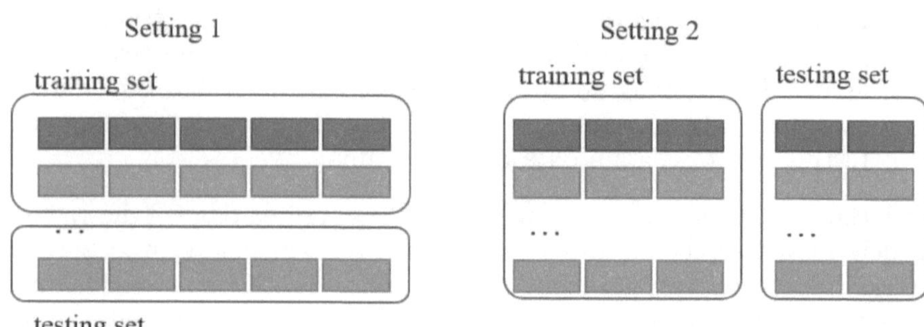

Figure 5.10: Splitting strategies for cross-validation: splitting by drivers (setting 1) and splitting by rounds (setting 2).

5.7.2 Results of Learning from Previous Maneuvers

The average F1 score and the accuracy of all four models on the validation sets are show in Table 5.1. Both variants that extract information from previous maneuver show significant improvements. The overall best score is obtained by G_{lstm}, which uses an LSTM layer for capturing M_{i-1}.

To find the optimal network, different configurations of the network are tested. The configuration of the network include the numbers of neurons for each layer and also the number of layers in each input branch. Different regularization parameters are also applied to each input path individually:

- Changing the dropout rate in range from 0.3 to 0.7

- Changing the standard deviation of noise that is used to data augmentation in range 0.05 to 0.2

In all settings, incorporating M_{i-1} always leads to improvement of the prediction performance.

The improvement observed for $G_s(S_i, M_{i-1})$ confirms that the statistical values of a maneuver execution can be used for describing the driving style. This is also the traditional approach for characterizing a driver. The additional improvement obtained by $G_{lstm}(S_i, M_{i-1})$ indicates that there is more information about the driver, which can be extracted by considering the whole progression of a maneuver execution instead of only using its statistical values.

Table 5.1: Mean of $F1$ and accuracy score on validation set over 10-fold cross-validation setting

Model	F1 Score	Accuracy
Threshold	81.9 %	89.8 %
$G(S_i)$	86.8 %	92.6 %
$G_s(S_i, M_{i-1})$	88.8 %	93.2 %
$G_{lstm}(S_i, M_{i-1})$	91.1 %	94.7 %

Evaluation Using Different Past Maneuvers

As mentioned in Section 5.3, our test route also consists of other maneuvers that can be used as input for the model. In this section, we evaluate and compare the impact of using two other left turn maneuvers as M_{i-1} in predicting taken gaps. The results are produced using the LSTM architecture in Figure 5.8 — $G_{lstm}(S_i, M_{i-1})$.

Table 5.2 shows the F1 and Accuracy scores on predicting the taken gap at main left turn scenario. Overall, the system still benefits the most when it extracts information from roundabout maneuvers. Since we are predicting taken gaps at a left turn maneuver, the first intuition would be the system should gain more information when we use other past left turn as input. However, we have to note that these two left turn maneuvers are located in a 30-zone with low traffic, whereas the main left turn is located at 50-zone with high-traffic roads. The behavior of the these two left turns are thus different from the main left turn. On the other hand, the roundabout maneuver is a complex one which is longer and require more inputs from driver, therefore there should be more information that

Table 5.2: Comparison on the impact of extracting information from different maneuvers to personalize the prediction of left turn decision

Used Maneuver	F1 Score	Accuracy
Left turn 1	90.2 %	93.7 %
Left turn 2	90.0 %	93.6 %
Roundabout	91.1 %	94.7 %

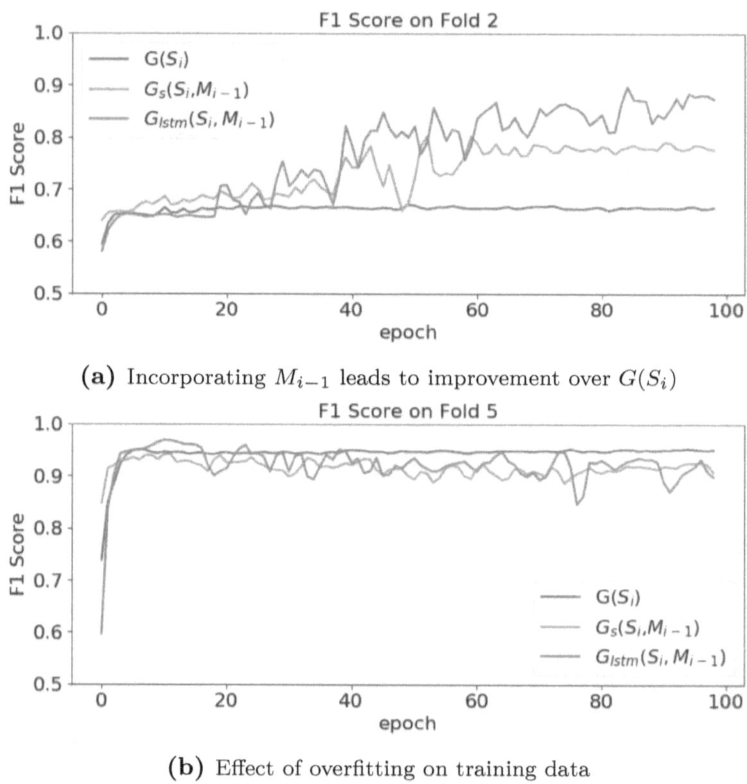

(a) Incorporating M_{i-1} leads to improvement over $G(S_i)$

(b) Effect of overfitting on training data

Figure 5.11: Learning curve of different validation folds over number of training epoch. Figure 5.11a shows the improvement of personalized models.

can be extract from a roundabout maneuver that is helpful for predicting taken gap

A Deeper Look at Each Training Fold

Reviewing the cross-validation results, we can observe the improvement of $G_s(S_i, M_{i-1})$ and $G_{lstm}(S_i, M_{i-1})$ over $G(S_i)$ in eight out of ten folds. The maximal improvement reaches 18.5 % in F1 score which translates to 9.4 % accuracy. Figure 5.11a shows the validation score of all three models over 100 epochs in one fold. We can see that $G(S_i)$ gets stuck and its best score is quite low, whereas, both G_s and G_{lstm} models benefit from the extra input M_{i-1} and reach much higher F1 score. Although the validation score is increasing over the epoch, we observe the first slight effect of overfitting as the score of both G_{lstm} and G_s keeps fluctuating. The problem is that, as mentioned in Section 5.5.3, the maneuver M_{i-1} contains not only information about driver but also about the previous situation S_{i-1}, which is not relevant for the current situation.

In Figure 5.11b, the effects of overfitting on both G_s and G_{lstm} can be further observed. It shows that the performances of both models are equal at the early state of the training process. After about 40 epochs, G_s and G_{lstm} start overfitting on the training data, which results in the drop of their performance on the validation set. Note that, the Figure 5.11 shows the complete 100 training epochs for visualizing the effects of overfitting. As described in Section 5.5.3, the actual training process is terminated by early stopping mechanism.

5.8 Conclusion

In this chapter,a new approach are introduced, that can learn the dependency between maneuver execution, namely to extract the information about driver behavior and style. This dependency between maneuver execution can be used to improve the performance of the prediction task in driver assistance systems. The first analysis on the clustering results shows that it is profitable using past maneuver to predict the impending left turn decision. Furthermore, we implement the proposed approach using neural networks as building blocks and empirically evaluate the model in a left-turn situation using on-road data. The results show that the model is able to extract the driver impact from the past maneuver executions and can use it improve the prediction quality by more than 9 % in terms of F1 score. The proposed architecture with two input paths to the network—one for the current maneuver and one for an abstraction of the recent history of past maneuvers—enables us to test different configurations for extracting driver information. We compared two approaches for extracting driver's information from maneuver execution, which make use of statistical features and an LSTM layer. The LSTM model shows improvement over a model that only uses statistical features and fully connected layers. It implies that we can extract more information about the driver than could be provided by using only statistical features of the raw maneuver data.

6 Driver Identification

Inter-personalization addresses the problem of identifying the differences between different users. To compare between inter- and intra-personalization, we can visualize, as shown in Figure 6.1, the driver behavior as a distribution over an imaginal 1-D driver behavior space ($\mathbf{x}$-axis). The inter personalization can then be seen as the differences between the means (μ_a, μ_b) of two driver behavior distributions. On the other hand, the intra-personalization is the differences between driver behaviors of a same driver (Δv). The problem of inter-personalization is therefore related to the problem of identifying driver based on behavioral data.

In this chapter, we introduce a framework to identify drivers based on driver behaviors captured in vehicle dynamic signals. Here the focus is laid on three aspects:

- Utilization of vehicle dynamics as inputs but not other signals that are usually used in identification tasks (i. e. finger print, facial recognition, etc.).

- Scalable approach that can be used to differentiate new drivers but not just a small set of driver in a given training set.

- The identification should be based on maneuver level. The aim is to be able to extract driver behaviors for identifying drivers but not focus on small microscopic changes in the input signals.

6.1 Introduction

Driver identification has been studied in different aspects. Most approaches focus on discovering features that can be used to improve the identification performance, often using machine learning methods. In this case, data from a number of drivers are recorded, and driver identification is then formulated as a classification problem where models are trained to distinguish between these drivers. While such approaches can learn to distinguish a set of given drivers, they can typically not extrapolate this knowledge to

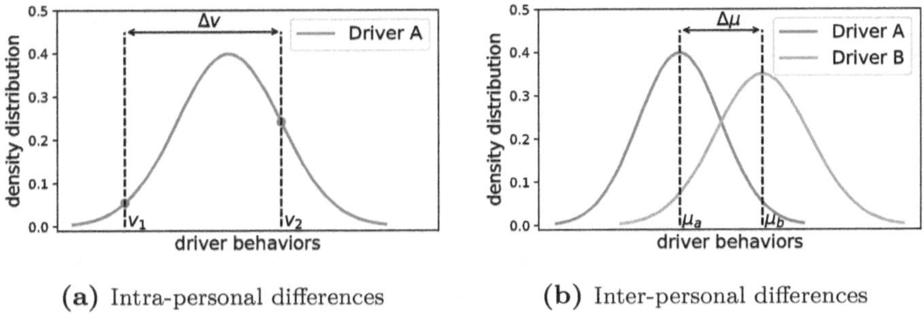

(a) Intra-personal differences **(b)** Inter-personal differences

Figure 6.1: Intra personalization can be viewed as the differences between driver behaviors of the same drivers (Δv). Inter personalization concentrates on the differences between drivers ($\Delta \mu$).

the recognition of new drivers. To overcome this restriction, it would be required to collect data and labels from the new drivers, as well as to re-train the model on the newly collected data, which is often infeasible in a practical setting. Thus, this type of techniques is not suitable to deal with the problem of new and changing drivers.

Although there are different techniques, that can be used to precisely identify a driver using a camera with face detection, voice recognition or even with special sensors such as finger print identification, driver identification based on the vehicle dynamics information is still a major research topic, because vehicle dynamics information are widely available and can be extracted from the CAN-Bus. So, there is no need to install additional sensors on the car. The second reason being that vehicle dynamics are closely related to the driving style. In fact, there are many works to determine driving style based on vehicle dynamics information (Vaitkus et al., 2014, Van Ly et al., 2013). Analyses on vehicle dynamics are therefore a step forward to understand the driving behavior and the driving style. This is also the foundation to build up personalized systems to better support drivers.

Maneuver-based approaches have been widely used in analyzing driving tasks. In Chapter 5, the dependency between roundabout and left-turn maneuvers is used to personalize the prediction. Hallac et al. (2016) present an approach to classify drivers using only left-turn maneuvers. The advantages of a maneuver-based approach is that it can focus on the behavior of the driver in a specific situation, whereas this information can get lost when analyzing a long chunk of driving data.

This chapter presents a novel approach on driver identification that can deal with the problem of new and changing drivers. The approach exclusively focuses on maneuver-based identification and uses only vehicle dynamics as features. In addition to driver identification, the approach can be used to further understand driving style and behaviors.

The remainder of this chapter is structured as follows. Section 6.2 formally defines the problem of driver identification as a comparison problem. Next, Section 6.3 presents the approach, which uses a Siamese neural network architecture in combination with long short-term memory (LSTM). In Section 6.4 the proposed approach is applied to a large data set and the results are analyzed along various dimensions. Further, it is scrutinized, whether the network based on the approach is able to extract driver behaviors from the input signals. Section 6.5 summarizes and concludes this chapter.

6.2 Problem Formulation

6.2.1 Driver Classification

The problem of driver identification is usually formulated as a classification problem. In other words, given a sample x, a model has to predict to which driver D_i does x belong to. The drivers are the classes and are fixed before the training and testing process. The output of such models are usually estimates $P(D_i \mid x)$ for the probability that the given input was generated by the given drivers. The final prediction is the class with highest probability

$$Y = \underset{D_i}{\mathrm{argmax}}\, P(D_i \mid x) \tag{6.1}$$

As mentioned above, the set of drivers is fixed. For example, if we have a model that has learned to discriminate between maneuvers of drivers D_1 and D_2, the model cannot be used to recognize maneuvers from a third driver D_3, but will instead assign them to either D_1 or D_2.

6.2.2 Driver Identification as Comparison Problem

In this chapter, we reformulate the problem of driver classification as a pairwise comparison problem. Instead of predicting whether some driver characteristics x belongs to a driver D, we predict whether two characteristics belong to the same driver or not. More precisely, given two samples

x_i and x_j which have been generated by drivers D_i and D_j respectively, we now have to learn a function f to predict whether $D_i = D_j$ or not, i.e., whether x_i and x_j have been generated by the same driver or not:

$$y_{i,j} = f(x_i, x_j) \text{ with } y_{i,j} = \begin{cases} 0 & \text{if } D_i \equiv D_j \\ 1 & \text{if } D_i \neq D_j \end{cases} \tag{6.2}$$

In general, x_i can be any property or characteristic for driver behavior. In this chapter, since our focus is laid on the vehicle dynamics of left-turn maneuvers, we will therefore the notation M instead of x. Formally, a maneuver M_i is a multivariate time series of vehicle dynamics generated by driver performing a left-turn maneuver.

$$M_i = (m_i^{(1)}, m_i^{(2)}, ..., m_i^{(n)}) \tag{6.3}$$

where $n > 0$ is the length of the time series, and $m_i^{(t)}$ is a column vector of sensor values at time step t. Assume that we have p sensors value at each time point t then $m_i^{(t)} \in \mathbb{R}^p$ and $M_i \in \mathbb{R}^{n \times p}$.

A simple use case for this would be the following scenario: a driver D_i uses the car on the first day and generates a set of maneuvers. The car stores these maneuvers with label D_i. On the next day, driver D_j uses the car. The car can then observe the maneuvers executed by this driver and run a comparison with the maneuvers recorded from D_i. It can then determine whether D_i and D_j are a same person or not.

This approach does not limit the number of drivers that can be identified since it does not learn to extract a specific feature for each predefined driver. It learns to extract individual features that are useful to distinguish between drivers. A new driver can be integrated into the system as a set of maneuver executed by that driver. The identification process will compare this maneuver set against a newly executed maneuver.

6.3 Driver Identification Using A Siamese LSTM Architecture

6.3.1 Siamese Architecture

Siamese networks have been successfully used for different identification tasks, mostly in visual computing domains such as face verification (Taigman et al., 2014), signature identification (Bromley et al., 1994) or one-shot image recognition (Koch et al., 2015). As illustrated in Figure 6.2, a

Siamese network consists of two identical encoders, which have the purpose of mapping the inputs into an embedding space while preserving the structural properties of interest. Formally, we can define the encoder as the encoding function $g(.)$ which maps each input M_i into an d-dimensional embedding vector E_i ($E_i \in \mathbb{R}^d$). Where E_i is in much smaller dimensional space than the original input maneuver M_i ($d << n \times p$).

$$E_i = g(M_i) \tag{6.4}$$

The outputs $g(M_i)$ and $g(M_j)$ of the two identical encoders are joined by a merge layer. This layer combines the embedding vectors and computes the final output of the Siamese network. There are two principled ways to design the merge layer. The first approach is to simply compute the distance between the embedding vectors and return it as output. The second approach is the use of further hidden layers to learn a function that can compare the input embedding vectors.

Let $d(.,.)$ be a distance function between two vectors. The comparison function from Equation (6.2) becomes:

$$\begin{aligned}
y_{i,j} &= f(M_i, M_j) \\
&= d(g(M_i), g(M_j)) \\
&= d(E_i, E_j)
\end{aligned} \tag{6.5}$$

In this chapter we use the distance function as the merge layer of two encoders, the reason being that the embedding vectors will be directly optimized based on the distance and could therefore be easier to interpret. Using a sophisticated merge layer will bind the comparison of the embedding vectors to the merge layer, and the embedding space will become a hidden representation that can only be used by the merge layer. In addition, a sophisticated merge layer will add more complexity to the model, which increases the risk of overfitting.

6.3.2 Loss Function

For each pair of maneuver executions (M_i, M_j), we have two possible labels indicating if they originate from the same person or not. Although this is similar to the binary classification problem, using a classification loss (such as binary cross-entropy) or a regression loss (such as mean squared error or mean absolute error) is not suitable to optimize the embedding network. The problem is that these loss functions are designed to optimize

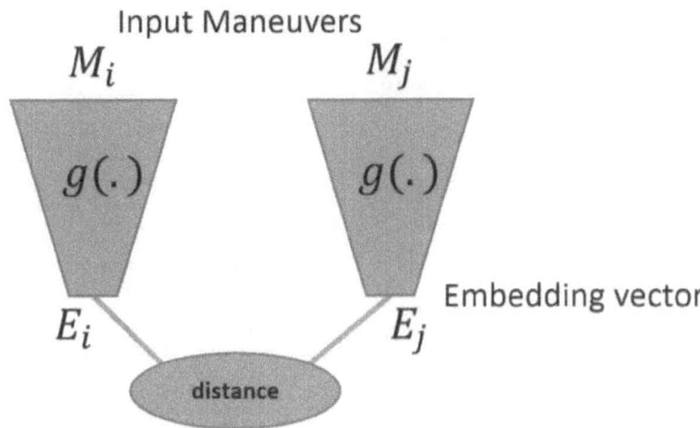

Figure 6.2: Siamese architecture to learn embedding vector from maneuver executions

the output (in this case, the distance d between $g(M_i)$ and $g(M_j)$) to become exactly the label. This means the distance between all dissimilar pairs of maneuver execution will be strictly forced to be 1. Therefore, a perfect network has to be able to map all maneuvers pairs from different drivers onto two points with the distance of exact 1. Theoretically, it is only possible to map $k+1$ different drivers into a k-dimensional space so that this requirement is fulfilled. However, to distinguish between drivers, it is enough to have the distance of all dissimilar pair to be larger than the distance between similar pairs. For this purpose the contrastive loss function is used to optimize the network.

The idea of contrastive loss function was described in (Hadsell et al., 2006). This loss function is similar to the cross entropy loss, except that it optimizes the distance between $g(x_1)$ and $g(x_2)$ instead of the prediction probability. In addition, the margin m is introduced to the loss. For computing the loss for a dissimilar pair, its distance will be clipped at m. Dissimilar pairs, whose distances are larger than or equal to m will not contribute to the loss. This loss function allows the network to learn to map the dissimilar pair to be be larger than or equal to m, not forcing it to be exactly m. The distance between maneuver executions from a same person will be optimized to be zero. Formally, the contrastive loss function is defined as:

$$L(Y, E_1, E_2) = \frac{1-Y}{2} d^2(E_1, E_2) + \frac{Y}{2} \max(0, m - d(E_1, E_2))^2 \quad (6.6)$$

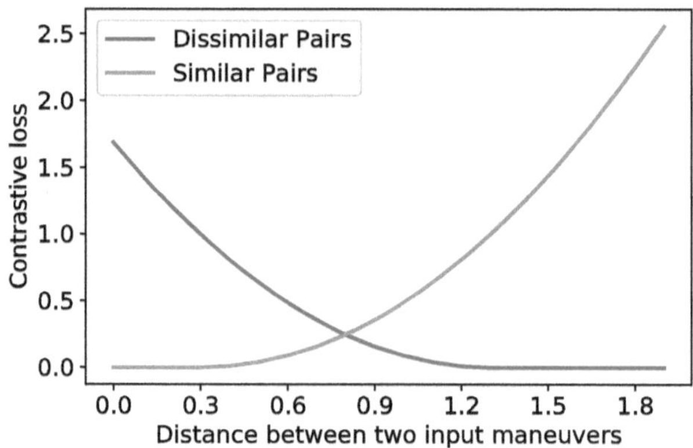

Figure 6.3

Here, Y is the label for the input pair (E_1, E_2): $Y = 0$ if E_1 and E_2 are generated by the same driver and $Y = 1$ otherwise. $d(\cdot, \cdot)$ is the distance function between two embedding vectors E_1 and E_2. Different types of distance functions can be used here, depending of the purpose of the task. Here, we can simply use the Euclidean distance:

$$d(E_i, E_j) = \|E_i - E_j\|_2 \tag{6.7}$$

In addition to the commonly used margin m, a lower margin δ is also introduced, which defines a lower threshold distances between maneuver executions from the same person. The loss function will optimize the network so that the distance of similar pairs will become smaller than or equal to the lower margin. Formally the resulting loss function is defined as:

$$L(Y, E_1, E_2) = \frac{1 - Y}{2} \max(d(E_1, E_2) - \delta, 0)^2$$
$$+ \frac{Y}{2} \max(0, m - d(E_1, E_2))^2 \tag{6.8}$$

This is a generalization of the contrastive loss function since with $\delta = 0$ it will become the original contrastive loss. Without the lower margin, the loss function will force the network to map all maneuvers of a driver to the same vector. In other words, a perfect model will have to be able to map all maneuvers of the same driver to a single point in the embedding space. The idea of using a lower margin is based on the fact that the driver style is composed of different factors and can change over time (slowly or fast). Maneuver executions generated by the same person are similar but not necessarily the exactly the same.

6.3.3 Embedding Network

The key component of the Siamese network architecture is the encoder function g. As it is shown in previous chapter, long short-term memory network (LSTM, cf. Section 2.2.2) is suitable for capturing time series of maneuver executions. We will therefore model the encoding function g using LSTM networks. In addition, the attention mechanism introduced by Bahdanau et al. (2015) is also incorporated in the networks. Recently, several works in the automotive domain also start using this type of network to capture time series data generated by vehicular sensors or to predict a driver's intentions (Hallac et al., 2018, Zyner et al., 2017).

To find the optimal network, a grid search is apply on different network configuration. The best architecture consists of 3 bi-directional LSTM layers stacked on top of each others. The last LSTM layer is followed by an attention layer as the last layer, output of the attention layer is also the output layer of encoder network. The number of hidden neurons in each of the single LSTM layers are 16, 32 and 32 units respectively. This results in 32, 64 and 64 hidden neurons in the bi-directional settings.

The attention layer is a special layer, which learns to evaluate the outputs of the LSTM layer at each time step. It decides how much the output of each time step will contribute to the final embedding vector. In this work, implementation of Raffel and Ellis (2016) is used:

$$E = \sum_{t=0}^{n} \mathbf{h}^{(t)} \alpha^{(t)} \text{ with } \sum_{t=0}^{n} \alpha^{(i)} = 1 \tag{6.9}$$

where $h^{(t)}$ is the output of the t-th time step of the last LSTM layer, $\alpha^{(t)}$ is the weighting factor for each time step, which is learned by the attention layer. Finally, E is the output of the embedding function. An illustration of the last LSTM layer and the attention mechanism in our architecture is shown on the left side of Figure 6.4. The right part of Figure 6.4 illustrates the construction of the proposed encoder, which includes 3 bidirectional LSTM layers and an attention layer stacked together.

The attention mechanism we used can be understood as a special pooling layer, which learns to weight each time step and accumulates them accordingly. The result of the attention layer (the last layer) is a 64-dimensional

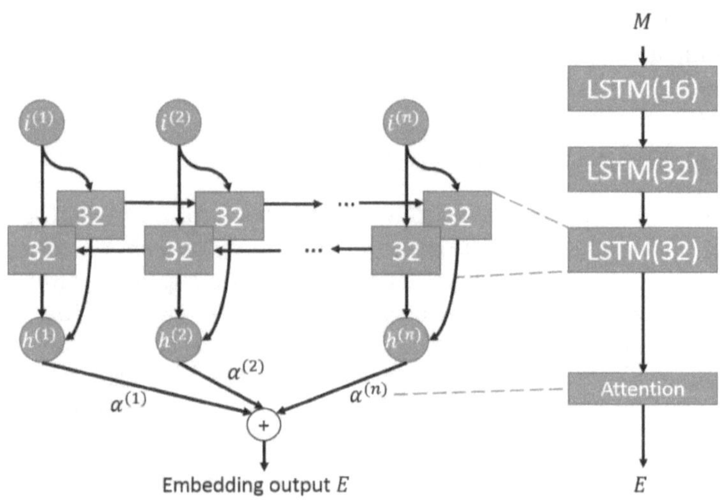

Figure 6.4: Illustration for the last bi-directional LSTM layer and the attention mechanism. $i^{(t)}$ show the input node which is the output of previous LSTM layer. $h^{(t)}$ is the combination of outputs from both direction of the bi-directional LSTM layer. $\alpha^{(t)}$ are the weights learned by attention layer

embedding vector E, which is the projection of the input M into the embedding space.

6.4 Evaluation

6.4.1 Evaluation Setup

The data set used in this work was collected from 32 drivers. The test drivers are chosen such that they cover a wide range of ages (see Table 6.1) and driving experiences (the annual mileage varies from few hundreds to 50,000 km). The route that was chosen for recording data is in an urban

Table 6.1: Age distribution of test drivers

Age	Number of drivers	%
≤ 26	10	31.2 %
$(27, 60)$	14	43.8 %
≥ 60	8	25.0 %

area with real-world traffic. Test drivers are asked to drive normally like their daily commute. The data set was collected with the PRORETA vehicle (cf. Section 3.3)

The left-turn maneuvers that are considered took place in a zone with a 30 km/h speed limit. This is accomplished using the map information, GPS and the steering wheel signal. From each maneuver, the vehicle dynamics information is extracted and fed as the input to the further processing steps. This signals include vehicle speed, longitudinal and lateral acceleration, steering wheel speed and yaw rate of the vehicle.

In total, 1920 left turn maneuvers were extracted which can be used to generate more than 1.84 million unique pairs of maneuver executions. A cross validation is run on the data set, for each fold two drivers are left out for testing and the other 30 drivers for training. In total, there are 496 unique pairs of drivers that can be used as a test set, which results in 496 folds for the cross-validation setting.

Since there are 30 drivers in each training set, the probability of sampling two maneuvers from the same driver is only $\frac{1}{30} \approx 3.3\,\%$. To deal with this class imbalance, down-sampling the negative pairs is applied so that the fraction of examples in the positive class is around 20–25 %. This sampling process is repeated before each training epoch in order to make sure that no data point gets wasted.

6.4.2 Driver Identification

ROC Analysis

First, it is tested whether the Siamese network is capable of differentiating a pair of maneuver executions from the same driver and that from different drivers. To that end, all maneuver pairs are ranked according to their distance as defined in Equation (6.7), and the area under the receiver operating characteristic (or AUC for short) curve is calculated as the metric to evaluate the results.

In general, the AUC metric estimates the probability that for a pair of a positive and a negative item, the positive is ranked before the negative item (Fawcett, 2006). In our case, positive items are maneuver pairs from the same driver, and negative items are maneuver pairs from different drivers. So given a two pairs of maneuvers, one from identical and one from different drivers, our AUC metric essentially estimates the probability that the distance between a maneuver pair from the same driver is lower than the distance between a maneuver pair from different drivers. An AUC of

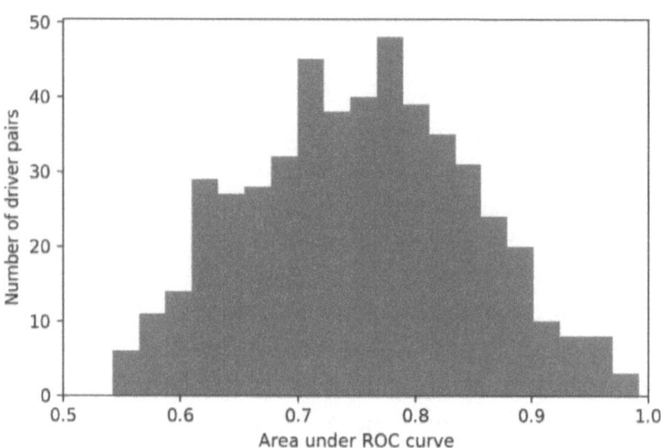

Figure 6.5: Histogram of area under ROC curve (AUC) computed from 496 driver pairs

0.5 means all maneuvers pairs are ranked randomly, and a value of 1.0 mean all pairs are perfectly ranked.

Figure 6.5 shows the histogram of AUC values over all 496 driver pairs in the test sets. The average AUC over all folds lies at 0.753. The driver pair that is the hardest to separate has an AUC of 0.55, and the score of the driver pair with the best separation lies at 0.99. This means that there are drivers which have very similar behavior and their maneuver executions can be hardly separated when only using vehicle dynamic information. Also an AUC of 0.99 shows that there are drivers whose maneuver execution can be almost perfectly distinguished by the network. This is an expected result due to the fact that drivers can have similar or dissimilar driving styles.

Visualization of the Embedding Space

A key advantage of embedding method is that it reduces the input to a 64-dimensional vector. Instead of storing the entire execution trace for all maneuvers M_i, we only have to store each maneuver's embedding vector $E_i = g(M_i)$. We can visualize these embeddings by further reducing the embedding vectors to a two-dimensional space. Several different techniques can be used for this purpose, including principal component analysis (PCA) or t-distributed stochastic neighbor embedding (t-SNE, cf. Section 2.1.2).

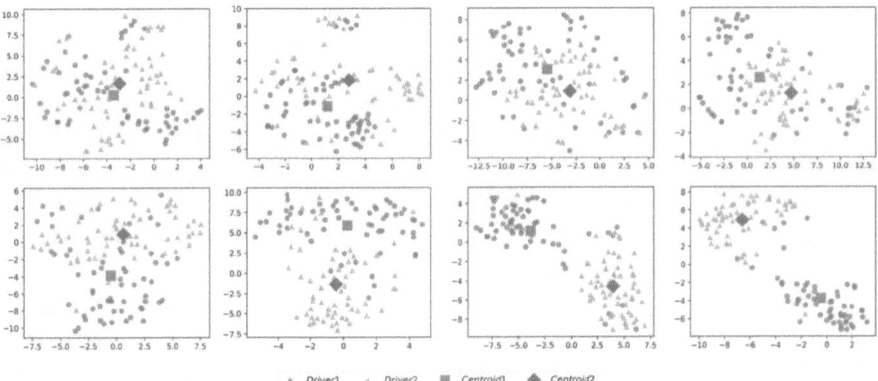

Figure 6.6: Comparison between t-SNE visualization of maneuver execution on the original feature space versus on the embedding space. The first row is the visualization of the original input, the second row shows the embedding space. Figures on the same column are computed from the same driver pair. The blue and orange points on each subfigure show the maneuver executions of each driver. The centroids are computed as the mean of all maneuver executions of each driver.

In this work we use the latter in order to compare the separation between selected driver pairs in the original and in the embedding space. Figure 6.6 shows the t-SNE visualization of four different driver pairs (the four columns) in the original input space (the first row) and in the embedding space (the second row). The driver pairs were selected based on their AUCs so that they span the separability range. The AUCs for the four driver pairs in the columns (in order from left to right) are 65.6 %, 74.5 %, 81.8 % and 91.9 % respectively. Each point in the plots corresponds to one maneuver, where maneuvers from the two drivers are distinguished with blue circles and red triangles respectively.

In the embedding space (second row), one can see that even with relatively low AUC, the maneuver executions form two fairly well separated clusters. With increasing AUCs, the separability of the driver pairs noticeably increases (from left to right), whereas in the original space (first row), no clear cluster separation is visible.

Finally, recall that the tested driver pairs have not been used for training the network. This shows that the ability to extract driver information from maneuver execution can be generalized to discriminating between unseen drivers.

6.4.3 Comparison between Sets of Maneuvers

Until now, we only evaluated the ability of the network to distinguish drivers given a single maneuver pair. However, the prediction performance can be further improved when comparing a set of maneuvers. With the assumption that the driver does not change during a driving session, we can easily collect a set of maneuvers that belongs to the same driver. For example, we can define a driving session as the period, where the driver's seat belt remains buckled up.

In the following, it can be shown that there is another advantage of the maneuver-based approach, namely that the prediction performance will increase along with the number of maneuvers that the driver has performed. Here we use the notation D_i^p to denote a set of p maneuvers that are generated by the driver i:

$$D_i^p = \{M_{i,1}, M_{i,2}, ..., M_{i,p}\} \tag{6.10}$$

To compare if D_i^p and D_j^q are generated by the same driver, we can actually compare $n = p \times q$ unique pairs of individual maneuver executions, and we can obtain a prediction for each one of these pairs. As long as the individual predictions are better than random, the accumulated prediction will increase along with the number of maneuver pair that the networks receive for comparison. Using a simple voting mechanism we only need to predict more than $\frac{n}{2}$ pairs correctly in order to output the correct identification. In fact, given the AUC for correctly predicting one pair of maneuvers is ρ, with $\rho > 0.5$, then the accumulated AUC when classifying one pair of maneuver set (D_i^p, D_j^q) is expected to be:

$$P(\rho, n) = 1 - \sum_{i=0}^{[n/2]} \binom{n}{k} \rho^k (1 - \rho)^{n-k} \tag{6.11}$$

with

$$\binom{n}{k} = \frac{n!}{k! \, (n - k)!} \tag{6.12}$$

The expected score can be reached if all maneuver pairs are independent. This concept is related to the ensemble technique deployed by random forest (Section 2.2.3). Instead of building multiple models to boost up the performance, we can take advantage of the data recording process to build up multiple maneuver pairs that have the same ground truth.

To verify if the Equation (6.11) actually holds, we compute the empirical ROC score in multi-maneuvers settings with $\rho = 0.7$. To do this, we need to firstly extract drivers pairs, which has the similar AUC score $\simeq 0.7$ when computed in the 1 maneuver setting (Section 6.4.2). There are about 30 driver pairs that satisfy this constraint (cf. Figure 6.5). For each $n = q \times p$ we randomly choose p and q unique maneuvers and use them to form a pair of two maneuver sets (D_i^p, D_j^q). We repeated this process $M = 400$ times to generate a data set which consists of M pairs of maneuver sets (D_i^p, D_j^q). About 50 % of the pairs are from the same driver. The AUC score is then computed on these data sets.

The blue line in Figure 6.7 shows the expected AUC score over the number of maneuver pairs (n) with ρ is set to 0.7. The orange line shows the average of the empirical AUC scores over the number of pairs. The light orange area show the standard deviation of the empirical AUC scores.

We can observe that there are 4 drops at $n = 7, 11, 13$ and 17. This can be explained since the value of n, in these cases, is a prime number and can only be factorized as a product of 1 and itself ($n = 1 \times n$). Which means n pairs are generated by a combination between a set of $p = 1$ maneuver (D_i^1) and a set of $q = n$ maneuvers (D_j^n). Maneuver pairs generated by these two sets are not truly independent since they always contains a same maneuver from D_i^1. The prediction performance, therefore, strongly depends on how well D_i^1 was chosen.

Another way to combine the information of multiple maneuvers of a single driver is to compute the average values of their embedding vectors, i.e., their cluster centroid. Figure 6.6 shows the centroids of the two drivers with a green square and a red diamond respectively. These centroids can be seen as a description of the driver. In the case of a clear separation, the task of identifying the driver can be reduced to compute the distance of the performed maneuvers to the existing centroids in embedding space. New identified maneuvers can be used to update the centroid of the driver. As can be seen in Figure 6.6, the distances between the centroids of each driver's maneuver increase with better separation in the embedding space (second row of graphs). Conversely, however, the distances between the centroids remain approximately the same in the original space. In fact, both centroids are always quite close to each other, which also illustrates the lack of separability in the original data representation.

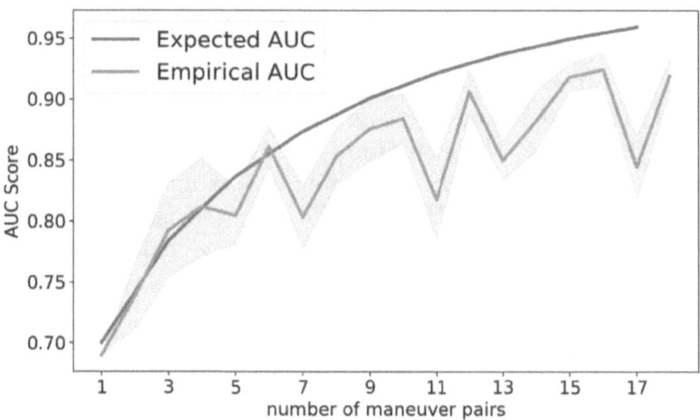

Figure 6.7: Area under ROC score (AUCs) of correctly classifying D_i^p against D_j^q over the number of seen maneuver pairs $n = p \times q$

6.4.4 Feature Importance Analysis

Until this point, we have showed the ability of the proposed network in identifying drivers based on vehicle dynamics. A requirement addressed at the beginning of the chapter is still remaining, whether the developed neural network is able to capture the driver behaviors to make predictions or if it just looks at a few data points (or peaks) from the input signals. The former case is a desirable property that we are aiming for when training the network. In contrast, the later case means that the network only extract small differences in the time series to make the decision. To answer this question, this section conducts a deeper analysis on the network components to provide a better understanding of how the network processes each maneuver.

To understand what kind of features is considered as important by the network we need to evaluate how each time step contributes to the prediction. By the fact that LSTMs use gate mechanism to weight the importance of different components (input, output and memory), we can evaluate the input gate at each time step to see how much the input contributes to the network. The values of the input gate indicate how importance is the input at a specific time step.

The input gates control how much a current time step contributes to the network (cf. Section 2.2.2). The values of input gate lay in the range

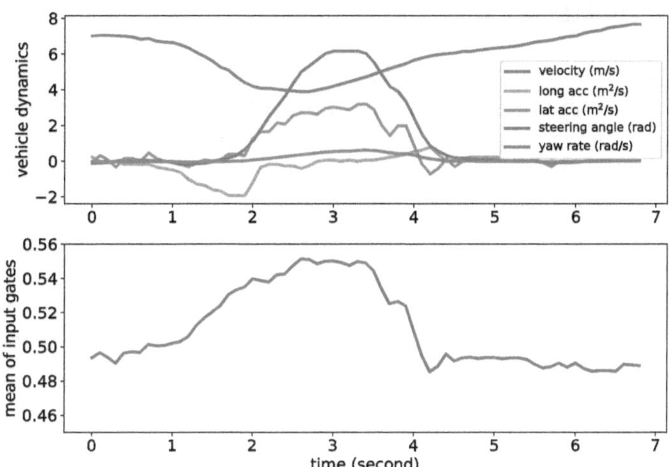

Figure 6.8: The first figure shows vehicle dynamics of a left turn maneuver in a 30 km/h zone. The second figure shows the mean of the activation function from input gates over each time step. The higher the value of input gate is, the more relevant information the time step contains. More visualizations of maneuvers from different drivers can be found in Appendix A

$[0, 1]$. 0 means that the time step is not relevant and is not used in the prediction. 1 means this time step is important for the prediction.

By visualizing the value of input gates at each time step, we can observe that the developed network does not look at only a few data points to make the final prediction. It rather considers the whole input sequence. The upper sub-figure of Figure 6.8 shows the sensor signals over time of a left turn maneuver. The lower sub-figure shows the activation of input gates, which indicate the importance of the corresponding time step.

Furthermore, the network put more weights the decelerating maneuver (negative longitudinal acceleration) and the actual turning maneuver (high lateral acceleration, steering angle and yaw rate). When the driver perform these maneuvers, the values of the input gates are relatively higher than at other time steps. This behaviors of the network are consistent and can be observed on all drivers (see Appendix A). This is a desirable property of the network since such maneuvers contains more information about driver behaviors.

6.5 Conclusion

In this chapter we showed how to reformulate the problem of driver identification as a comparison problem. This allows to not only assign driving maneuvers to a set of previously observed drivers, but can also generalize to new drivers in the sense that the learner can recognize whether two maneuvers are from the same, possibly previously unseen driver.

To this end, a Siamese architecture is used in combination with long short-term memory networks to demonstrate the ability to distinguish drivers by only using vehicle dynamics information. The approach is evaluated on unseen driver pairs. The results show that the proposed method is able to distinguish drivers with different driving styles. The distance between maneuvers can also be used as a scalar to measure the similarity between drivers.

The deeper analysis on the network components shows that the developed model is able to extract relevant maneuvers from the input time series and use them as basis for distinguish drivers. We also show that there is the potential of using multiple maneuvers to even increase the identification performance. The more maneuvers the model observes from a driver, the better it can distinguish her from others.

Furthermore, the resulting embedding space is enriched with driver information and are suitable for visualizing the differences between drivers. This can be use as a tool for examine the difference between driver types.

7 Summary and Outlook

In the last few years, advanced driver assistance systems (ADAS) have been gaining more importance. They provide drivers with driving comfort and safety in various situations, e.g. parking, blind spot warnings, collision avoidance systems, adaptive cruise control, etc. Traffic accidents and many adverse consequences can be precluded by deploying ADAS. To improve the performance of ADAS, it is essential to take driver's data into account. The ability to distill valuable information from these data can help close the gap between a standard assistance system and the individual need of a driver. This will in turn increase user's acceptance of the ADAS. With that in mind, this book was conducted with the goal to build an adaptive personalized driver assistance system that can extract and incorporate driver's information into its predictions as well as recommendations.

In the scope of this book, we addressed four key research questions (cf. Chapter 1), which deeply delve into the problems of personalization in ADAS:

1. **How to anticipate driving maneuvers by analyzing the driving data?**
   To answer this question, we conducted an extensive analysis on predicting lane change maneuvers using various machine learning models. Our proposed model using long short-term memory (LSTM) neural network archived the best result in comparison with the other models. We also re-formulate the problem of predicting lane change to predicting the exact time left until the vehicle crossed the lane marking. We call it time-to-lane-change or TTLC for short. Although predicting TTLC is a harder problem, our LSTM network provides even slightly better results than in classification setting

2. **How to extract driver's information from driving data?**
   To learn about a driver we have to analyze the driving data and extract the information associated with driving style, preferences, skills etc. However extracting that information is a complex and challenging task. Considering a driving trip as a whole fails to distinguish the

actions caused by external factors (e.g. road topography, oncoming traffic, etc.) and the intention of the driver. Therefore, the driving data is broken down into separated maneuvers. This helps change the problem of mining the driver's information from a long time series into mining the dependency between maneuver executions.

It is showed that driver's information can then be extracted by using either statistical properties of the maneuver executions or sophisticated methods for capturing the whole maneuver executions like neural network or LSTM (see Section 5.4, Section 5.6 and Chapter 6)

3. **How to deal with intra-personalization problem?**
   This research question aims at identifying the changing in the driving style of a driver between trips e.g. driving-to-work versus driving-to-vacation. That means the information derived from previous trips may not be appropriate for the current one. Assistance systems need to recognize the driving style online during the trip. Assuming that the driver does not change her style rapidly during a short time window of a trip, two consecutive maneuvers are most likely to be executed in the same driving style. The prediction of the next maneuver can be adjusted by incorporating input from the last maneuver. This process can be explained as: by observing how driver performs the past maneuvers, one can draw an inference about how she would perform the next maneuver. With that, our system does not need to observe a maneuver multiple times before delivering a recommendation.

   As stated in the second question, there are many external factors affecting the behavior of the driver. Since we want to address the changes in the driving style of the driver, it is necessary to remove the impact of the external factors in the driving data before predicting her behavioral modification. Chapter 5 formally describes how to perform these steps and adapt the prediction for the next maneuver based on the previous one.

   The evaluation showed that complicated maneuvers tend to contain more relevant information of driver than simpler ones. This is an innovative approach to measure the amount of individual information contained in each type of maneuver execution.

4. **How to determine and visualize the inter-individual differences between drivers?**

The answer of the previous research question shows that it is possible to extract individual information from driving data, which can be used for personalizing a driver behavior model. The fourth question dives into mining driver information from driving data that does not change over longer period. To tackle this problem, Chapter 6 evaluates the ability of learning models in distinguishing different drivers by inspecting the vehicle dynamics. This directly addresses the problem of inter-personalization in driver behavior. In particular, a Siamese network is designed which aims at mapping maneuver executions into a new dimensional space. The network is trained such that maneuvers executed by the same driver are mapped close together while maneuvers by different drivers are mapped significantly further from each other. The results show that the proposed network is able to extract the driver information that did not change or marginally changed over the longer time period (in this case 3-4 hours).

The learned embedding space can be leveraged to visualize the similarity of drivers given their maneuver executions. In this way, the inter-individual differences between drivers can visually observed.

Machine learning has been playing a central role in many current state of the art researches in personalized driver assistance systems. Among others, there are two main reasons for this trend in the development of ADASs: the increasing amount of data and the advances of machine learning methods. These two reasons will not become obsolete in the near future. They will keep being the driving force for research and adoption of machine learning in personalization of ADASs.

The focus of this **book** lies on extracting possible information about driver from the available behavioral data, which are in most cases vehicle dynamics. The human factors involved in the driving tasks are however complex. Aspects like strong emotional reactions may prompt abrupt changes in driving behaviors and cannot be detected on time. The employment of further sensors can provide additional information to the assistance systems. For examples, the information about heart rates or the interactions between the driver and the other passengers can be used to improve the personalized recommendations.

Another challenge of personalization is the so-called filter b ubble effect. It describes the effect w hen u sers r eceive p ersonalized recommendations without noticing that they are personalized. The users are therefore trapped in a bubble that blocks out information or perception about the world.

This problem can be alleviated by actively showing users what part of the recommendation is personalized and what would be a standard recommendation. For example, in the left turn use case, the personalized recommendation can be also visualized as the differences between a current recommended gap size and a standard gap size. Another way to dealing with the filter bubble effect is to be more transparent about what the underlining algorithm "thinks". To this end, results of algorithms like driving style clustering are suitable to be provided as explanations. The ability to reveal the driver with inside information will increase the transparency. The driver becomes more aware of how the algorithm produces specific personalized recommendations. This approach was also experimentally implemented in PRORETA 4 vehicle and was positively perceived by testers. However there is a need for further studies on this subject.